THE RAY SOCIETY

INSTITUTED 1844

This volume is No. 154 of the series

LONDON

JOHN RAY

DICTIONARIOLUM

TRILINGUE

EDITIO PRIMA

1675

Facsimile with an Introduction
by
WILLIAM T. STEARN

LONDON
THE RAY SOCIETY
1981

MADE AND PRINTED IN GREAT BRITAIN
BY UNWIN BROTHERS LTD.
THE GRESHAM PRESS, OLD WOKING, SURREY

SET IN PHOTOTYPESETTING, GARAMOND

ISBN No. 0 903874 16 4

Available from:

Publications Sales,
British Museum (Natural History),
Cromwell Road,
London SW7 5BD,
England.

TO

SIR GEOFFREY KEYNES

MA., M.D., LL.D., D.LITT.,
F.R.C.P., F.R.C.S. F.B.A.

in appreciation of
his catholic interests
versatility and scholarship
notably exemplified in his
bibliographies of

WILLIAM HARVEY
THOMAS BROWNE
ROBERT HOOKE
AND
JOHN RAY

the definitive biography of
WILLIAM HARVEY
and other works

John Ray's *Dictionariolum trilingue* a forgotten vocabulary

William T. Stearn

Few copies of small much-used school-books survive the constant and sometimes rough handling which they customarily receive. Worn and soiled, with pages torn, scribbled upon or lost, with sentences or words underlined, their scholastic duties over, they end their service by being discarded and replaced by later editions or by later rival works. Thus there are few copies now extant of the first edition of Benjamin Hall Kennedy's *The Revised Latin Primer* (1888; re-issued up to 1962) and even fewer of his *Elementary Latin Grammar* (1862), certainly not in mint condition. The first edition of William Cobbett's *A Grammar of the English Language . . . for the Use of Soldiers, Sailors, Apprentices and Plough-boys* (1819) is a very rare little book, yet its continual re-issue as *Cobbett's Easy Grammar* down to 1958 proves its long-lasting utility. The first edition of John Ray's once popular *Dictionariolum trilingue; secundum Locos communes Nominibus usitatioribus Anglicis, Latinis, Graecis, Ordine παραλλήλως [parallelos], dispositis* (1675), being of a like unpretentious nature, has likewise suffered virtual extinction. Its former utility is attested by the issue of a second edition in 1685, a third in 1696 as *Nomenclator classicus, sive Dictionariolum trilingue*, a fourth in 1703, a fifth in 1706, a sixth in 1717, a seventh in 1726 and an eighth in 1736. Having thus served generation after generation of schoolboys for over sixty years, it then passed into obscurity. The total of recorded surviving copies of all editions in the British Isles is only 22; six of these belong to the first edition now reproduced by the Ray Society. No longer serviceable in schools, nevertheless the *Dictionariolum* possesses enough historical value for British naturalists, by recording English vernacular names (with their Latin and Greek equivalents) used in the 17th century for birds, fish, insects, mammals and plants, to justify its reprinting. It has also a more general interest since Ray also listed clothes, diseases, drinks, domestic utensils, foods, measures and much else under his 32 headings, thereby indicating what evidently had importance, utility or interest in his day.

According to Ray's friend William Derham (1657–1735), whose *Select Remains of the late learned John Ray, M.A. and F.R.S., with his Life* (posthumously published in 1760) was reprinted by the Ray Society in 1846, the *Dictionariolum* (called *Nomenclator classicus* from 1696 onwards) owed its origin to the needs of Ray's young pupils, Francis and Thomas Willughby, the sons of his deceased friend Sir Francis Willughby. Thomas Willughby later became Lord Middleton. To quote Derham, 'these two gentlemen being then very young (the eldest not four years of age), Mr. Ray, as a faithful trustee, betook himself to the instruction of them; and Mrs. Ray herself also (after their marriage) was an assistant in this matter, she being the person that taught Lord Middleton his letters, and to read English. For the sake and service of these two young gentlemen, Mr. Ray composed his *Nomenclator Classicus,* which was first published in this very year 1672. And the reason why he composed this book, when many others were extant, was because there were multitudes of errors in all Nomenclatures then in use, especially in the names of animals and plants. And as he was the best able of any man living to assign the true meaning of both the Latin and Greek names, so it was a very useful and valuable task he undertook; serviceable, not only to school-boys, but to the amendment of our Dictionaries and Lexicons, as may be seen in some of the best of them, that have been published since that time. In most of which, I observe, that they make use of the significations of words assigned by Mr. Willughby and Mr. Ray, which scarce ever was done before by the old grammarians; but yet sometimes they cannot forbear approving of the old signification, which inveterate custom hath made familiar to them'. Derham erred, however, in attributing the first publication of the *Dictionariolum* to 1672 instead of 1675.

Ray's Career

John Ray (1627–1705) was the supreme British naturalist of the 17th century, the author of numerous works relating not only to botany and zoology but also to travel, theology and the English language, a man both erudite and many-sided and possessed moreover of an attractive, modest, genial and generous personality. The massive biography by Charles E. Raven (1885–1964) brings together virtually everything relevant to his career. As a Cambridge theologian, a diligent many-sided scholar and an enthusiastic amateur naturalist with a remarkably extensive firsthand knowledge of British plants, birds and insects, Raven was well fitted to portray sympathetically the life and to assess authoritatively the achievements of Ray with whom he had so much in common. His *John Ray, Naturalist* (1942) was appropriately dedicated 'to all who like John Ray have sacrificed security and careers for conscience' sake'. As Keynes has remarked, this biography of Ray 'must have opened the eyes of many besides myself to the extraordinary qualities and endearing character of that remarkable man'. Shorter accounts have been provided by Arber (1943), Crowther (1960), Stearn (1973), Webster (1975) and others. Keynes (1951, 1976) gives a detailed description of his numerous works, bringing forward 'evidence in a bibliographical dress of his modesty, his loyalty and his integrity'.

Ray was born on 29 November 1627 at the smithy of Black Notley, near Braintree, Essex, eastern England, where his father was the village blacksmith. A stone monument, erected at the initiative of the Bishop of London, Henry Compton (1632–1713), himself a keen botanist and gardener, stands in the churchyard at Black Notley. It has a long Latin inscription typical of the period but far more important monuments are the works of scholarship which gained him international renown during his lifetime and which remain of interest and relevance even today.

Evidently through the influence of a local clergyman who recognized the boy's ability and promise, Ray became a pupil in 1638 at the Braintree Grammar School. Being of humble rural origin, he would never have made his way into the world of learning, would never have been able to use his talents over so wide a field of scholarship and would never have left posterity so much indebted to him but for the good basic schooling he received at Braintree and for a scholarship at the University of Cambridge arising out of Braintree rents. He entered Cambridge in 1644 and later became a lecturer in Greek, mathematics and humanities. He began the study of plants growing around Cambridge in 1650 when recovering from an illness, his delight in their beauty and diversity leading him ultimately to produce his *Catalogus Plantarum circa Cantabrigiam nascentium* (1660), of which an English translation by A. H. Ewen and C. T. Prime was published in 1975. Between 1658 and 1662 he travelled extensively in England and Wales and visited the Isle of Man and southern Scotland. The pernicious Act of Uniformity in 1662 caused him to forfeit his Fellowship at Trinity College, Cambridge on grounds of conscience and end both his ecclesiastical and university careers. He was now thirty-five and financially distressed, as were some other 2000 clergymen of the Church of England, by sacrifice of employment. His friend and former student Francis Willughby (1635–1672) came to his aid and together they travelled on the Continent of Europe from 1663 to 1665, as related in Ray's *Observations topographical, moral and physiological; made on a Journey through Part of the Low-Countries, Germany, Italy, and France* (1673).

Willughby died on 3 July 1672, aged only thirty-seven, bequeathing Ray an annuity of £60 a year, adequate for his modest needs; during the next three and a half years Ray stayed at Middleton, as tutor to Willughby's sons Francis and Thomas and in 1673 he married Margaret Oakley. In 1679 Ray and his wife settled down at Black Notley; here their three daughters were born; here he lived, investigating and writing and from 1685 onwards suffering much ill-health, up to his death on 17 January 1705.

At Black Notley Ray prepared the major works, *Historia Plantarum* (3 volumes, 1686–1704), *Synopsis methodica Stirpium Britannicarum* (1690), *Miscellaneous Discourses concerning the Dissolution and Changes of the World* (1692), *Synopsis methodica Animalium Quadrupedum et Serpenti Generis* (1693), *Stirpium Europaearum extra Britannias nascentium* (1694), *Historia Insectorum* (1710) and *Synopsis Avium et Piscium* (1713), on which his reputation as an encyclopaedic naturalist rests. Ray's interests ranged, however, far beyond natural history. Thus he published in 1670 *A Collection of English Proverbs* (2nd ed., 1678; 3rd ed. 1737), in 1674 *A Collection of English Words not generally used . . . and an Account of the Preparing and Refining such Metals and Minerals as are gotten in England* (2nd ed., 1691), in 1691 *The Wisdom of God manifested in the Works of the Creation* (2nd ed. 1692; 3rd ed. 1701; 4th ed. 1704; 5th ed. 1709, which was also published in French in 1714 and in German in 1717) and in 1700 *Persuasive to a Holy Life.*

As stated by Raven (1942: 453) *The Wisdom of God* 'is certainly his most popular and influential achievement. Published as a slim octavo volume of 249 pages, in a first edition of 500 copies, it was reprinted in a second edition of 382 pages in 1692, in a third of 414 pages in 1701 and in a fourth of 464 pages in 1704. It was reissued many times during the next century; it formed the basis of Derham's Boyle Lectures in 1711–12; it supplied the background for the thoughts of Gilbert White and indeed for the naturalists of three generations; it was imitated, and extensively plagiarised, by Paley in his famous *Natural Theology* [1802]; and more than any other single book it initiates the true adventure of modern science, and is the ancestor of the *Origin of Species* [1859] and of *L'Evolution Créatice* [1907]'.

The botanic gardens at Oxford, founded in 1621, Edinburgh, founded in 1670, and Chelsea, founded in 1673, existed in Ray's time but such active centres of botanical taxonomic research as the British Museum (Natural History) and the Royal Botanic Gardens, Kew had yet to be created. When Ray returned in 1679 to Black Notley, Hans Sloane (1660–1753) was but a young medical student, just arrived out of Ulster and far from being the owner of the vast collections which in 1753 formed the nucleus of the British Museum. Thus Ray's home at Black Notley became botanically the 17th-century equivalent of the British Museum and Kew. To his obscure Essex village came letters and information about plants gladly sent from the Philippines by a Jesuit pharmacist, Georg Joseph Kamel, from Virginia by an Anglican missionary, John Banister, and from numerous correspondents in the British Isles eager that their findings could be incorporated in his monumental works. These contemporaries esteemed Ray as the scholar who could best appreciate their endeavours and who would honestly record his indebtedness to their co-operation. Thus to and from Black Notley, there stretched many far-flung lines of friendship in the promotion of learning among lovers of nature.

The Dictionariolum Trilingue

Compared with Ray's major works listed above, his *Dictionariolum trilingue* (literally 'Little three-language dictionary') is certainly humble and seemingly insignificant, but the same regard to the public benefit and interest motivated its publication and the same width of scholarship underlay it as for these other works. Ray and Willughby had naturally to give much attention to the application and origin of the ancient classical names used by their predecessors and interpreted by their contemporaries. Ray in particular considered many contemporary interpretations to be mistaken or misleading. To remedy this he compiled his *Dictionariolum* and stated his reasons in the preface: 'Having lately had occasion to review some of the last Published *English* and *Latin* Nomenclatures, I observed in them some inveterate Errors, especially in the names of the Animals and Plants still continued, which I thought it might not be amiss to correct; notwithstanding that for their Antiquity, they may plead Prescription, and for their Universality in our Schools, general Approbation. For certainly better it is that Children, when they first learn the *Latin* Tongue, be taught the true names of things than false or wrong ones, which must again afterwards with a double labour be unlearned, and in the mean time hinder the true understanding of Authors, and occasion other mistakes: not to say that it is some discredit to our *English* Schools in general, to retein and teach their Youth what is manifestly erroneous. Such Mistakes are the rendring of a Grasshopper in Latin *Cicada*, a Lapwing *Upupa*, a Caterpillar *Gryllus*, a Barbel *Mullus*, and many others, which the Reader may find corrected in this Book.'

He confessed that 'in the names of Apparel, several sorts of Viands, parts of Buildings, Utensils and Implements of Household stuff, Instruments and Tools of Husbandry and Gardening, I have not, nor indeed can I fully satisfie my self. That which I chiefly minded was to correct such manifest Mistakes as the little insight I have in the History of Plants and Animals inabled me to discern in former

Nomenclatures, which I have accordingly done in this little Work, following therein the Judgment of the best Authors of those several Histories.'

From the natural history standpoint the most relevant and interesting are the sections on 'Stones and Metals' (pp.5–6), 'Parts & Adjuncts of Plants' (pp.6–8), 'Herbs' (pp.8–14), 'Trees and Shrubs' (pp.15–19), 'proper Parts & Adjuncts of Animals' (pp.19–21) i.e. of fish (pp.19–20), birds (p.20) and mammals (p.21), 'four-footed Beasts' (pp.21–25), 'Birds' (pp.25–29), 'Fishes' (pp.29–32), 'Insects' (pp.33–34). More interesting from a medical standpoint are the sections on 'parts of Mans Body' (pp.34–38), 'Some Accidents of the Body' (pp.39–40) and 'Diseases' (pp.41–44).

It is of interest to note both how few have been the changes in the commonly used English vernacular names for plants and animals adopted by Ray in 1675 and, despite the drastic revisions of their nomenclature and classification by Carl Linnaeus in the 18th century, how many of the Latin names have passed into modern scientific nomenclature, most of them being adopted as specific epithets by Linnaeus. Thus Ray's lists have a familiar look which makes discrepancies all the more apparent. 'Primrose' has here *Primula veris* as its Latin equivalent, rightly so, for this species (now *Primula vulgaris* Hudson), being 'the firstling of Spring', was so designated by pre-Linnaean authors, but the name *Primula veris* L. is now restricted to the 'Cowslips or Paigles' for which Ray gave the Latin equivalent *Paralysis* following earlier pre-Linnaean usage; thus John Parkinson in his *Paradisi in Sole Paradisus terrestris* 243 (1629) headed his chapter 35 '*Primula veris* & *Paralysis*. Primroses and Cowslips.' As Parkinson said in explanation of the name *Primula*, 'they shew by their flowring the new Spring to bee comming on, they being as it were the first Embassadours thereof.' For *Primula auricula* L. and its hybrids now called 'Auriculas' Ray had the entry 'Bears-ear. Auricula ursi', likewise in accordance with earlier usage exemplified in Parkinson's chapter heading (p.235) '*Auricula Ursi*. Beares eares.' There are indeed various names in Ray's list of plants now obsolete or little used but apparently then standard, e.g. 'Bluebottle. Cyanus' (*Centaurea cyanus* L.), 'Flower de luce. Iris', 'Jack in the hedge. Alliaria', 'Maudlin-tansie. Ageratum' (*Achillea ageratum* L.) and 'Sage of Jerusalem. Pulmonaria'. They are however, few.

Ray added comments on some plant names. Thus he noted under 'Asparagus or Sperage. Asparagus' that 'This also is by the vulgar corruptly called Sparrow-grass', under 'Marigold. Calendula' that 'This is thought to be the *Caltha* of the Poets'. He gave a warning under 'Laurel. Laurocerasus' (*Prunus laurocerasus* L., cherry laurel) against confusion with the bay laurel or true laurel (*Laurus nobilis* L.), a warning still necessary in view of the poisonous nature of leaves of the first and the culinary value of those of the second: 'The Tree we now commonly call Laurel, is not the Laurel or *Laurus* of the Ancients, that being our Bay-tree, but a Plant unknown till of late, bearing an esculent Fruit like a Cherry, and yet an ever-green. This is to be carefully heeded, lest any one be deceived by the confusion of these names'. Earlier in the 17th century *Prunus laurocerasus*, a native of northern Asia Minor and the adjacent eastern Balkan Peninsula, was still a rare plant not long introduced from Istanbul and enthusiastically described at length by Parkinson in 1629 as 'this beautifull Bay' with 'fruits as large or great as Flanders Cherries, many growing one by another on a long stalke, as the flowers did, which are very blacke and shining on the outside, with a little point at the end, and reasonable sweete in taste' as he had observed 'by them I had of Master Iames Cole a Merchant of London lately deceased, which grew at his house in Highgate, where there is a faire tree which hee defended from the bitternesse of the weather in winter by casting a blanket over the toppe thereof every yeare, thereby the better to preserve it'. Ray's comment indicates that by 1676 it had become well known.

Ray's vernacular names for birds, fish and mammals are likewise mostly those in use today, with their Latin equivalents likewise mostly preserved as generic names or specific epithets, but here again there are discrepancies. He had no entry under 'Rabbit' for the common rabbit (*Oryctolagus cuniculus, Lepus cuniculus* L.) as at this time the original use of the word 'rabbit' only for the young of the species still continued; George Turberville writing on falconry in 1575 stated that 'the hare and the conie are called in their first yeare leverets and rabets'. Ray's entry reads 'A Coney. Cuniculus.' In view of the importance of the horse, there are eight entries for different kinds. As with plants Ray provided some explanatory notes: 'A Bat though it flies, hath no affinity to Birds, not so much as a flying Serpent: and though it be not properly a Quadruped, yet hath it Claws in the Wings, which answer to Forelegs'. The entry 'A Dormouse. Glis' has likewise a note 'That little Beast which with us is usually called a Dormouse is not *Glis*, but *Mus avellarum* of Naturalists. The *Glis* is unknown to me, and therefore because it is vulgarly taken for our Dormouse, I have let it stand.' The native British dormouse, the hazel dormouse, apparently unknown to the Romans, is now known scientifically as *Muscardinus avellanarius* (L.). The *Glis* of the Romans, unknown to Ray, is the much larger fat or edible dormouse (*Glis glis* (L.)), now naturalized in England, the bodies of which, according to an old Roman cookery book (cf. Flower & Rosenbaum, 1958), should be stuffed with minced pork and dormouse flesh, seasoned with pepper, pine-kernels etc., and cooked in an oven. Linnaeus classified the hazel dormouse as a mouse (*Mus*) and the fat dormouse as a squirrel (*Sciurus*).

The difference between the faunas of southern and northern Europe and hence the lack of definitely applicable classical names for some fish and birds likely to be familiar to British schoolboys caused Ray much difficulty. Thus to justify his entry 'A Lapwing. Vanellus' relating to the green plover or lapwing (*Vanellus vanellus* (L.)) he added: '*Vanellus* is a new-made name of the French *Vanneau*. This bird by a great mistake hath been generally taken to be the *upupa* of the Antients, which is now by all acknowledged to be the Hoopo.' He noted under 'A Linnet. Linaria' that 'This is also no antient Latin word, but a newly imposed name, as is also *Gallinula* for a More-hen, *Pluvialis* for a Plover, *Ruticilla* for a Redstart, *Rubecula* for a Robin red-breast'. At this time *Regulus*, 'found with us in England', 'hath no English name that I know'; it has subsequently become well known as the goldcrest (*Regulus regulus* (L.)). 'The Salmon being a Fish proper to the Ocean was probably unknown to the antient Greeks, and therefore hath no name in that language.' Various of the Greek names taken by Ray from ancient Greek survive unchanged in modern Greek.

Ray's sectional heading 10 'Of Fishes' merely conformed to the popular application of the noun 'fish' to almost any animal living wholly in water, ranging from shellfish (such as cockles, limpets, mussels and oysters), cephalopods (such as cuttlefish), and crustaceans (such as crabs, crayfish, lobsters and shrimps), to dolphins, porpoises and whales, which are still legally 'royal fish' or 'fishes royal' in Britain, as well as true fish. Ray as a naturalist was, of course, well aware of the artificiality of including such diverse creatures under the one designation 'Fishes'; as a lexicographer he knew that this was the heading under which the general public would seek them in a work intended for general use. His list summarized the results of much painstaking enquiry among fishermen and fishmongers and of careful study of the works of erudite predecessors, notably Conrad Gessner, Pierre Belon, Ulisse Aldrovandi, Guillaume Rondelet and Hippolyto Salviani. This was backed by a first-hand knowledge of the fish themselves. Ray's interest in fishes went back, according to Raven, at least to the year 1658. In 1661 he recorded the fish seen in 'the market at Scarborough on the Yorkshire coast and in 1673 he published a list of fishes taken at Penzance, Cornwall, provided by 'one of the ancientest and most experienced fishermen', and another list of the freshwater fish found in England. In December 1674 he was beginning to write the massive work ultimately published as *Francisci Willughbeii Armig. De Historia Piscium Libri Quatuor* (1686). Such a background of enquiry gives Ray's lists and notes an authority standing far above that of a simple compiler. Most of the vernacular names he recorded in 1676 remain in use today, as they had been for many centuries before 1676. Of Ray's English names for 53 kinds of true fishes 48 are given as their present names in such a standard work as Alwyne Wheeler's *Key to the Fishes of Northern Europe* (1978), apart from minor differences in spelling, e.g. 'doree' (now 'dory'), 'loche' (now 'loach'), 'plaise' (now 'plaice'), 'roche' (now 'roach'). The majority of their Latin equivalents, as with other organisms, have passed into modern scientific nomenclature, e.g. *Pungitius, Barbus, Cyprinus, Torpedo, Leuciscus, Raja, Thymallus, Lampetra, Perca.* There have, however, been changes; thus the long slender fish with a long tapering snout listed by Ray as 'An Horn-fish or Needle-fish. Acus' is now called 'garfish' but its Greek equivalent βελονη (bĕlŏnē) has provided the specific name *Belone belone.*

It will be evident from the above, as also for the numerous other annotations scattered through the *Dictionariolum*, that this was far from being a simple compilation but was, on the contrary, despite its concise and unpretentious nature, a work based on profound and extensive erudition such as probably no other scholar of the period could have made. In many ways, quite apart from natural history, it reflects the life of the times by providing lists of words for which Latin equivalents would be useful. Thus there are consecutive entries for 'A Jail or Prison', 'A House of Correction', 'A Pair of Stocks', 'A Pillory', 'A Gallows'. More than 80 entries refer to ailments. As Raven states, 'indeed in days when Latin was still the common speech of educated men this little vocabulary in three columns, English Latin and Greek . . . must have been indispensable'. Its many editions are evidence of contemporary appreciation. Even today medical men, social, military, naval and agricultural historians, architects, anglers, theologians and mineralogists, as well as naturalists and latinists, can all find something of interest here.

Editions of the *Dictionariolum*

All the works of John Ray are described exhaustively in Geoffrey Keynes, *John Ray, a Bibliography* (London, 1951), of which a second edition entitled *John Ray, 1627–1705; a Bibliography, 1660–1970* (Amsterdam) was published in 1976. The following accounts of the editions of the *Dictionariolum* are quoted below from this second edition by gracious permission of the author.

Abbreviations for libraries holding copies of the *Dictionariolum* are: BLO (Bodleian Library, Oxford), BM (British Library, Bloomsbury, London), CLO (Codrington Library, Oxford), GUL (Glasgow University Library and Hunterian Collection), NLI (National Library of Ireland, Dublin), NLS (National Library of Scotland, Edinburgh), RSL (Royal Society, London), TCC (Trinity College, Cambridge).

In relating how he came to compile this massive contribution to Raian scholarship Sir Geoffrey Keynes remarked that 'Had Ray's work been restricted to botany, or had his character been less attractive, it may be that I should never have attempted to compile a full-scale bibliography of his works, but the versatility of his attainments, and the real nobility of his character made him irresistible.'

26. DICTIONARIOLUM TRILINGUE 8° 1675

Title, in double lines: Dictionariolum Trilingue: Secundum Locos Communes, Nominibus usitatioribus Anglicis, Latinis, Graecis, Ordine παραλλήλως dispositis. [*rule*] Opera Joannis Raii, M.A. Et Societatis Regiae Sodalis. [*double rule*]

Londini: Typis Andreae Clark, impensis Thomae Burrel, ad Insigne Pilae auratae sub Æde S. Dunstani in vico vulgò vocato Fleetstreet. 1675.

Collation: A–M^4; 48 leaves.

Contents: AI title; A2*a–b* *The Preface* signed *Jo. Ray*; A3*a*–M4*a* (pp.1–91) text: M4*b* blank. *Errata* at bottom of M4*a*.

Copies: BLO, TCC, RSL, NLS.

27. DICTIONARIOLUM TRILINGUE Second Edition. 8° 1685

Title, within double lines: Dictionariolum Trilingue: Secundum Locos Communes, Nominibus usitatioribus Anglicis, Latinis, Graecis, Ordine παραλλήλως dispositis. [*rule*] Opera Joannis Raii, M.A. Et Societatis Regiae Sodalis. [*rule*] Editio Altera. [*rule*]

Londini: Typis M. C. Impensis Christoph. Wilkinson, & Benj. Tooke, apud quos prostant è regione Ædis S. Dunstani in Vico vocato Fleetstreet, & in Coemeterio D. Pauli, 1685.

Collation: A^2 A–I^4 K^2; 40 leaves.

Contents: AI title; A2*a–b* *The Preface*, signed *Jo. Ray*; AI*a*–I4*b* (pp.1–70 [should be 72]) *Dictionariolum Trilingue*; K1*a*–K2*b* (pp.71–4 [should be 73–6]) *Facetiae quedam*.

Note: H4 and I1 are both paginated 63, 64.

Copy: BM (626.b.4(2)).

28. NOMENCLATOR NOVUS TRILINGUIS
Second Edition, second issue. 8° 1688

Title: Nomenclator Novus Trilinguis . . . Opera Joannis Raii . . . [*rule*] Editio Altera [*rule*]

Londini: Typis M. C. Impensis Christop. Wilkinson & Benj. Tooke. Vaeneunt apud Sam. Crouch in viculo Popes-Head Alley dicto è regione Cambii Regalis. 1688.

Collation, contents: As in no. 27, with cancel title.

Copy: CLO.

29. NOMENCLATOR CLASSICUS Third Edition. 8° 1696

Title, within double lines: Nomenclator Classicus, sive Dictionariolum Trilingue, Secundùm Locos communes, Nominibus usitatioribus Anglicis, Latinis, Graecis, ordine παραλλήλως dispositis. A Classical Nomenclator with The Gender and Declension of each Word and the Quantities of the Syllables. [*rule*] By John Ray, M.A. and Fellow of the Royal Society. [*rule*] To which are added Paradigmata of all the Declensions, as well Greek as Latin; with a Century of Proverbs, and a Collection of modest Jests in English, Latin and Greek, from good Authors. [*rule*] The Third Edition, carefully revised and corrected, containing many Hundreds of Words more than any Book of this Nature. For the Use of Schools. [*rule*]

London: Printed by Benj. Motte, and are to be Sold by the Booksellers of London and Westminster. 1696.

Collation: A^2 A–E^8 K^2; 44 leaves.

Contents: A1 title, A2*a–b The Preface*; A1*a*–K2*b* (pp.1–84) text.

Note: B2 has sign. C2, and the pagination at the end is irregular. It is correct up to p.78. The numbers then run 67, no number, 71–4.

Copy: G. L. Keynes.

29*a*. NOMENCLATOR CLASSICUS

Fourth edition, variant title. 8° 1703

Title: Nomenclator Classicus . . .]*&c., as in no. 29*]

London: Printed by Benj. Motte, for Sam. Smith and Benj. Walford, at the Prince's Arms in St. Paul's Church yard, 1696.

Collation, contents: as in no. 29.

Copy: V. A. Eyles.

30. NOMENCLATOR CLASSICUS Fourth edition. 8° 1703

Title within double rules: Nomenclator Classicus, sive Dictionariolum Trilingue, . . . [etc. as in no. 29] The Fourth Edition, . . . [etc. as in no. 29]

London: Printed by Benj. Motte, for John Taylor, at the Ship in S. Paul's Church-yard. 1703.

Collation, contents: As in no. 29, with the pagination corrected.

Note: This line-by-line reprint of no. 29 was announced for publication in the *Term Catalogue* for February 1703.

Copy: G. L. Keynes.

31. NOMENCLATOR CLASSICUS Fifth Edition. 8° 1706

Title, within double lines: Nomenclator Classicus, sive Dictionariolum Trilingue, . . . The Fifth Edition, . . . [*rule*]
London: Printed by Benj. Motte, for John Taylor, at the Ship in St. Paul's Church-yard. 1706.

Collation: A² B–D, D–E⁸ K²; 44 leaves (sign. D repeated).

Contents: A1 title; A2*a–b* *The Preface*; B1*a*–K2*b* (pp.1–84) text.

Copies: BM (12933.bb.4(1)), GUL, G. L. Keynes.

32. NOMENCLATOR CLASSICUS 'Fourth' Edition. 8° 1708

Title, within double lines: Nomenclator Classicus, sive Dictionariolum Trilingue . . . To which is added . . . [*rule*] The Fourth Edition, . . . [*rule*]
London: Printed by Benj. Motte, and Re-printed by E. Waters in School-House-Lane. Dublin, 1708.

Collation: A–L⁴; 44 leaves.

Contents: A1 title; A2*a–b* *The Preface*; A3*a*–L4*b* (pp.1–84) text.

Copy: NLI.

33. NOMENCLATOR CLASSICUS 'Fifth' Edition. 12° 1715

Title: Nomenclator Classicus, sive Dictionariolum. Secundùm Locos communes Nominibus usitatioribus Anglicis Latinis, ordine dispositis. A Classical Nomenclator, with The Gender and Declension of each Word, and the Quantities of the Syllables. [*rule*] By J. Ray, M.A. and

Fellow of the Royal Society. [*rule*] To which are added, Paradigmata of all the Declensions of the Latin; with a Century of Proverbs, and a Collection of modest Jests in English and Latin, from good Authors. [*rule*] The Fifth Edition, carefully Revised and Corrected, containing many hundred of Words more than any Book of this Nature. For the Use of Schools. [*rule*]

Dublin: Printed by A. Rhames, for Eliphal Dobson, at the Stationer's-Arms in Castle-Street, 1715.

Collation: A–H^6; 48 leaves.

Contents: A1 title; A2*a*–H6*b* (pp.3–96) text. (The last five pages are unnumbered.)

Note: Ray's preface and all the Greek portions are omitted.

Copies: BM (12934.aa.17), LSL, NLI.

34. NOMENCLATOR CLASSICUS Sixth Edition. 8° 1717

Title, within double lines: Nomenclator Classicus, sive Dictionariolum Trilingue, . . . The Sixth Edition, . . .

London: Printed by Benj. Motte, for J. Taylor, at the Ship in Paternoster-Row, and Benj. Tooke at the Mid-dle Temple-Gate. 1717.

Collation: A^2 B–D, D, F–L^4 M^2; 44 leaves (sign. D repeated, sign. E omitted).

Contents: A1 title; A2*a–b* *The Preface*; B1*a*–M2*b* (pp.1–84) text.

Copies: BM (12923.aa.13), G. L. Keynes.

35. NOMENCLATOR CLASSICUS Seventh Edition. 8° 1726

Title, within double lines: Nomenclator Classicus, sive Dictionariolum Trilingue, . . . The Seventh Edition, . . .

London: Printed by D. Leach, for W. and J. Innys in St. Paul's Church-yard, B. Motte in Fleetstreet, and J. Clarke in Duck-Lane near West Smith-field. MDCCXXVI.

Collation: A^2 B–L^4 M^2; 44 leaves.

Contents: A1 title; A2*a*–*b* *The Preface*; B1*a*–M2*b* (pp.1–84) text.

Copies: BM (B.693.(1)), AUL, G. L. Keynes.

36. NOMENCLATOR CLASSICUS 'Eighth' Edition. 12° 1735

Title: Nomenclator Classicus, sive Dictionariolum. . . . To which are added, Paradigmata . . . As Also, A Collection of common Verbs and Ad-jectives in English and Latin, never before published. [*rule*] The eight Edition, carefully revised and corrected, Fitted for the Use of Schools. [*rule*]
Dublin: Printed and Sold by Luke Dillon, at the Bible in High-Street, MDCCXXXV.

Collation: A–M^6; 72 leaves.

Contents: A1 title; A2*a*–K3*a* (pp.3–113) Ray's text; K3*a*–M6*b* (pp.113–44) *Collectio Verborum Vulgariorum Anglo-Latina.*

Note: Ray's text ends in the middle of p.113. The rest of the book is occupied by the additions by an unknown hand announced on the title-page.

Copy: BM (12934.aa.15), G. L. Keynes.

37. NOMENCLATOR CLASSICUS Eighth Edition. 8° 1736

Title: Nomenclator Classicus, sive Dictionariolum Trilingue, . . . The Eighth Edition . . . [*rule*] For the Use of Schools. [*double rule*]
London, Printed by T. Wood, For W. Innys and R. Manby, at the West-End of St. Paul's, B. Motte in Fleet-street, and J. Clarke in Duck-lane. [*short rule*] M.DCC.XXXVI.

Collation: A–L^4; 44 leaves.

Contents: A1 title; A2*a*–*b* (pp.iii–iv) *The Preface*; A3*a*–L4*b* (pp.5–88) text.

Note: The copy collated is bound up with other pamphlets in a volume carrying a donor's book-plate of Francis Wrangham, 1842.

Copy: TCC.

Acknowledgements

Grateful thanks are due to a distinguished member of the Ray Society for generously making available his private copy of the first edition of the *Dictionariolum* reproduced here in facsimile, to Mr Peter Green (BM (NH)) for his care in photographing this, and to Sir Geoffrey Keynes for kind permission to reprint above his detailed bibliography of editions of the *Dictionariolum*.

It should be noted that for convenience of consultation the present facsimile has been printed slightly larger than the original, which has a print line of 9·7cm.

Sources of further information

ARBER, A. 1943. A seventeenth-century naturalist: John Ray. *Isis* 34:319–324.

CROWTHER, J. G. 1960. *Founders of British Science.* London (Cresset Press).

FLOWER, B. and ROSENBAUM, E. 1958. *The Roman Cookery Book, a critical Translation of the Art of Cooking by Apicius.* London (Harrap).

JACKSON, CHRISTINE E. 1968. *British Names of Birds.* London (Witherby).

KEYNES, G. L. 1951. *John Ray: a Bibliography.* London (Faber & Faber).

KEYNES, G. L. 1976. *John Ray, 1627–1705; a Bibliography, 1660–1970.* Amsterdam (Gerard Th. van Heusden).

POTTER, S. and SARGENT, L. 1973. *Pedigree: Essays on the Etymology of Words from Nature.* London (Collins).

PRIOR, R. C. A. 1879. *On the popular Names of British Plants.* 3rd ed. London.

RAVEN, C. E. 1942. *John Ray, Naturalist, his Life and Works.* Cambridge (Cambridge University Press).

STEARN, W. T. 1965. William Turner's 'Libellus', 1538, and 'Names of Herbes', 1548. Prefixed to Ray Society facsimile of Turner, *Libellus. The Names of Herbes.* London (Ray Society).

STEARN, W. T. 1973. Ray, Dillenius, Linnaeus and the 'Synopsis methodica Stirpium Britannicarum'. Prefixed to Ray Society facsimile of Ray, *Synopsis methodica.* London (Ray Society).

SWANN, H. KIRKE. 1912. *A Dictionary of English and Folk Names of British Birds, with their History, Meaning and first Usage.* London.

WEBSTER, C. 1975. Ray, John. In C. C. Gillispie (Ed.), *Dictionary of Scientific Biography* 11: 313–318. New York (Charles Scribner's Sons).

Appendix

INSCRIPTION ON THE MONUMENT
TO JOHN RAY IN THE CHURCHYARD OF BLACK NOTLEY

Eruditissimi Viri JOHANNIS RAIJ, A.M.
Quicquid mortale fuit,
Hoc in angusto tumulo reconditum est.
At *Scripta*
Non una continet Regio:
Et Fama undequaque celeberrima
Vetat Mori.
Collegii S. S. Trinitatis Cantab. fuit olim Socius,
Necnon Societatis Regiæ apud Londinenses Sodalis,
Egregium utriusque Ornamentum.
In omni Scientiarum genere
Tam Divinarum quam Humanarum
Versatissimus.
Et sicut alter Solomon (cui forfán Unico Secundus)
A Cedro ad Hyssopum,
Ab Animalium maximis, ad minima usque Insecta,
Exquisitam nactus est Notitiam.
Nec de Plantis solùm, quà patet Terræ facies
Accuratissimè disseruit,
Sed et intima ipsius viscera sagacissimè rimatus,
Quicquid notatu dignum in universa Naturâ
Descripsit.
Apud exteras Gentes agens,
Quæ aliorum Oculos fugerent, diligentèr exploravit,
Multaque scitu dignissima primus in Lucem protulit:
Quod superest, eâ Morum Simplicitate præditus,
Ut fuerit absque Invidia Doctus:
Sublimis Ingenii,
Et, quod raró accidit, demissi simul animi et modesti;
Non Sanguine et Genere insignis,
Sed quod majus
Propria Virtute Illustris.
De Opibus Titulisque obtinendis
Parum solicitus,
Hæc potius mereri voluit quam adipisci:
Dum sub Privato Lare, suâ Sorte contentus
(Fortunâ lautiori dignus) consenuit.
In Rebus aliis sibi modum facilè imposuit,
In Studiis nullum.
Quid Plura?
Hisce omnibus
Pietatem minimè fucatam adjunxit,
Ecclesiæ Anglicanæ
(Id quod supremo halitu confirmavit)
Totus et ex Animo addictus.
Sic benè latuit, benè vixit Vir beatus,
Quem Præsens Ætas colit, Postera mirabitur.

The Ray Society

Its history and aims

The Ray Society was founded in 1844 by a group of British naturalists which included Thomas Bell, J. S. Bowerbank, Edward Forbes, William Jardine, George Johnston, Edwin Lankester and Richard Owen. Its purpose, as they stated then, was 'the promotion of Natural History by the printing of original works in Zoology and Botany; of new editions of works of established merit; of rare Tracts and MSS; and of translations and reprints of foreign works; which are generally inaccessible'. The publication of learned books on natural history, with special but not exclusive relevance to the British fauna and flora, remains the object of the Society. It commemorates, with singular aptness, the great English naturalist John Ray (1627–1705). The founders evidently and rightly considered that Ray's breadth of interests, his learning and his eminence in all he touched provided ideals for emulation by the Society.

The publishing activities of the Ray Society began in 1845 with the issue of *Reports on the Progress of Zoology and Botany*, the first part of *A Monograph of the British Nudibranchiate Mollusca* by Alder and Hancock and a volume *On the Alternation of Generations* by Steenstrup, followed in 1846 by *Memorials of John Ray, Outlines of the Geography of Plants* by Meyen and *The Organization of Trilobites* by Burmeister. The diversity of subjects thus early established as coming within the scope of the Society has been maintained in subsequent volumes, of which many have become classics. Some of these important works would never have been written, let alone published with so many illustrations, had not the Ray Society undertaken their issue. Further details will be found in *The Ray Society, a bibliographical History* (1954) by R. Curle.

The Society's business is conducted by a Council of botanists and zoologists; this is elected at the Annual General Meeting and consists of a President, six Vice-Presidents, an Honorary Treasurer, an Honorary Secretary, an Honorary Foreign Secretary and twelve Councillors.

Membership of the Society is open to any person willing by subscription to promote its work of publishing scholarly contributions to British natural history, which are issued as funds and manuscripts permit.

Officers and Council of the Ray Society, 1980–81

Dictionariolum TRILINGUE:

SECUNDUM

LOCOS COMMUNES,

Nominibus usitatioribus

ANGLICIS, LATINIS, GRÆCIS,

Ordine παραλλήλως dispositis.

OPERA

JOANNIS RAII, M.A.

Et Societatis Regiæ Sodalis.

LONDINI:

Typis *Andreæ Clark*, impensis *Thomæ Burrel*, ad Insigne Pilæ auratæ sub Æde S. *Dunstani* in vico vulgò vocato Fleetstreet. 1675.

THE PREFACE.

HAving lately had occasion to review some of the last Published English and Latin Nomenclatures, I observed in them some inveterate Errors, especially in the names of Animals and Plants still continued, which I thought it might not be amiss to correct; notwithstanding that for their Antiquity, they may plead Prescription, and for their Universality in our Schools, general Approbation. For certainly better it is that Children, when they first learn the Latin Tongue, be taught the true names of things than false or wrong ones, which must again afterwards with a double labour be unlearned, and in the mean time hinder the true understanding of Authors, and occasion other mistakes: not to say that it is some discredit to our English Schools in general, to retein and teach their Youth what is manifestly erroneous. Such Mistakes are the rendring of a Grass-hopper in Latin Cicada, a Lapwing Upupa, a Caterpillar Gryllus, a Barbel Mullus, and many others, which the Reader may find corrected in this Book.

The Preface.

In the names of Apparel, ſeveral ſorts of Viands, parts of Buildings, Utenſils and Implements of Houſhold ſtuff, Inſtruments and Tools of Husbandry and Gardening, I have not, nor indeed can I fully ſatisfie my ſelf. Many of the names of ſuch things not having been by any good Author ſufficiently explained, or the things they ſignifie deſcribed. Beſides, the Antients uſed many Diſhes of meat, Inſtruments and Utenſils of Houſhold-ſtuff, &c. *which we uſe not, or different from what we now uſe; and many new ones there are uſed by us, which they knew not. Moreover it may be obſerved, that the ſame name is by divers Writers variouſly expounded, as indeed the Faſhions of things daily altering, and the names ſtill remaining as well with them as with us, they muſt needs be: So that all that can heer be done, is conveniently to accommodate names ſo as the things ſignified by them in each Language, may come as near and anſwer as well one the other as may be. I confeſs I have not heerin performed what I might have done, had I made it my deſign of a long time, though I hope in this reſpect the Reader will find ſome amendment. That which I chiefly minded was to correct ſuch manifeſt Miſtakes as the little inſight I have in the Hiſtory of Plants and Animals inabled me to diſcern in former Nomenclatures, which I have accordingly done in this little Work, following therein the Judgment of the beſt Authors of thoſe ſeveral Hiſtories.*

JO. RAY.

Dictionariolum TRILINGUE.

I.

Of Heaven.	De Cælo.		Περὶ Οὐρανοῦ.
The Skie	Æther, ĕris, *m.*	5	Αἰθὴρ, έρος, m.
A Star	Stella, æ, *f.*	5	Ἀστὴρ, έρος, m.
A Constellation	Sidus, ĕris, *n.*	3	Ἄστρον, ου, n.
A Comet or Blazing-star	Comēta, æ, *m.*	1	Κομήτης, ου, m.
A Planet	Planēta, æ, *m.*	1	Πλανήτης, ου, m.
Saturn	Saturnus, i, *m.*	3	Κρόνος, ου, m.
Jupiter	Jupĭter, Jovis, *m.*	5	Ζεὺς, Διός, m.
Mars	Mars, Martis, *m.*	5	Ἄρης, εος, m.
The Sun	Sol, solis, *m.*	3	Ἥλιος, ίου, m.
* *The Morning-Star*	Venus, ĕris, *f.*	2	Ἀφροδίτη, ης, f.
Mercury	Mercurius, ii, *m.*	1	Ἑρμῆς, οῦ, m.
The Moon	Luna, æ, *f.*	2	Σελήνη, ης, f.
New Moon	Novilunium, ii, *n.*	2	Νουμηνία, ας, f.
Half Moon	Luna semiplena.	3	Σελήνη διχότομος, ου, f.
Full Moon	Plenilunium, ii, *n.*	3	Πανσέληνον, ήνου, n.

* The Planet *Venus*, when it appears in the morning before Sun-rising, is called the Morning-Star, in Latin and Greek *Phosphorus*: when it appears in the Evening after Sun-set, it is called the Evening-Star, in Latin *Hesperus*.

A Sun-

A Sun-beam	Radius,ii, *m.*	5 Ἀκτὶν,ῖνος, f.
An Eclipse	Eclipsis,is, *f.*	2 c. Ἔκλειψις,εως, f.
The East	Oriens,tis, *m.*	2 Ἀνατολὴ,ῆς, f.
The West	Occidens,tis, *m.*	2 c. Δύσις,εως, f.
The North	Septentrio,ōnis, *m.*	3 Ἄρκτος,ου, com.
The South	Meridies,ei, *m.*	2 Μεσημβρεία,ας, f.

II.

Of the Elements and Meteors.	*De Elementis & Meteoris.*	Περὶ τῶν Στοιχείων καὶ Μετεώρων.
FIRE	Ignis,is, *m.*	5 Πῦρ,ρὸς, n.
Flame	Flamma,æ, *f.*	5 Φλὸξ,γὸς, f.
A Spark	Scintilla,æ, *f.*	5 Σπινθὴρ,ῆρος, m.
A Firebrand	Torris,is, *m.*	3 Δαλὸς,οῦ, m.
A Brand quenched	Titio,ōnis, *m.*	5 Θυμάλωψ,πος, m.
A live Coal	Pruna,æ, *f.*	5 Ἀνθρακὶς,ίδος, f.
A dead Coal	Carbo,ōnis, *m.*	5 Ἄνθραξ,ακος, m.
Heat	Calor,ōris, *m.*	5 Θερμότης,ητος, f.
Warmth	Tepor,ōris, *m.*	5 Χλιαρότης,τητος, f.
Burning	Incendium,ii, *n.*	3 Ἐμπρησμὸς,οῦ, m.
Brightness	Splendor,ōris, *m.*	2 Μαρμαρυγὴ,ῆς, f.
Smoak	Fumus,i, *m.*	3 Καπνὸς,οῦ, m.
Soot	Fulīgo,ĭnis, *f.*	5 Λιγνὺς,ύος, f. Ἄσβολος,όλου.
Embers	Favilla,æ, *f.*	2 Μαρίλα,ης, f.
Ashes	Cinis,ĕris, *m.*	3 Σποδὸς,οῦ, f.
Fuel	Fomes,ĭtis, *m.*	3 Πυρεῖον,ου, n.
AIR	Aer,ĕris, *m.*	5 Ἀὴρ,έρος, m.
A Cloud	Nubes,is, *f.*	1 c. Νέφος,εος,ους, n.
A Mist	Nebŭla,æ, *f.*	2 Ὁμίχλη,ης, f.
A Shower	Imber,bris, *m.*	3 Ὄμβρος,ου, m.
Rain	Pluvia,æ, *f.*	3 Ὑετὸς,οῦ, m.

The

The Rainbow	Iris, ĭdis, *f.*	5 Ἶρις, ιδος, f.
A Drop	Gutta, æ, *f.*	5 Σταγὼν, όνος, f.
Hail	Grando, ĭnis, *f.*	2 Χάλαζα, άζης, f.
Snow	Nix, nivis, *f.*	5 Χιὼν, όνος, f.
Dew	Ros, roris, *m.*	3 Δρόσος, ου, m.
Frost	Gelu, *indec. n.*	3 Πάγος, ου, m.
Hoar-frost	Pruīna, æ, *f.*	2 Πάχνη, ης, f.
Ice	Glacies, ei, *f.*	3 Κρύσταλλος, ου, m.
An Ice-cicle	Stiria, æ, *f.*	3 Σταλαγμὸς, οῦ, m.
Thunder	Tonĭtru, *indec. n.*	2 Βροντὴ, ῆς, f.
A Thunderbolt	Fulmen, ĭnis, *n.*	3 Κεραυνὸς, οῦ, m.
Lightning	Fulgur, ŭris, *n.*	2 Ἀστραπὴ, ῆς, f.
A Flash	Coruscatio, ōnis, *f.*	5 Λαμπηδὼν, όνος, f.
A Storm	Procella, æ, *f.*	2 Θύελλα, ης, f.
A Tempest	Tempestas, ātis, *f.*	5 Χειμὼν, ῶνος, m.
A Whirlwind	Turbo, ĭnis, *m.*	5 Λαίλαψ, απος, f.
A gentle wind	Aura, æ, *f.*	2 Αὔρα, ας, f.
The wind	Ventus, i, *m.*	3 Ἄνεμος, έμου, m.
The East-wind	Eurus, i, *m.*	1 Ἀπηλιώτης, ώτου, m.
The North-wind	Aquĭlo, ōnis, *m.*	1 Βορέας, έου, m.
The South-wind	Auster, ri, *m.*	3 Νότος, ου, m.
The West-wind	Zephyrus, i, *m.*	3 Ζέφυρος, ύρου, m.
Fair Weather	Serenĭtas, ātis, *f.*	2 Αἰθεία, ας, f.
Calm Weather	Tranquillĭtas, ātis, *f.*	2 Νηνεμία, ας, f.
WATER	Aqua, æ, *f.*	5 Ὕδωρ, ατος, n.
A Bubble	Bulla, æ, *f.*	5 Πομφόλυξ, υγος, f.
A well	Puteus, i, *m.*	5 Φρέαρ, ατος, n.
A Fountain or Spring	Fons, tis, *m.*	2 Πηγὴ, ῆς, f.
A Stream	Fluentum, i, *n.*	3 Ῥεῖθρον, ου, n.
A River	Fluvius, ii, *m.*	3 Ποταμὸς, οῦ, m.
A Rivulet	Rivus, i, *m.*	5 Ῥύαξ, ακος, m.
A Brook	Torrens, tis, *m.*	3 Χείμαῤῥος, άῤῥου, m.
The Bank of a River	Ripa, æ, *f.*	2 Ὄχθη, ης, f.
The Chanel	Alveus, i, *m.*	3 Ὀχετὸς, οῦ, m.
A Pool or Pond	Stagnum, i, *n.*	2 Λίμνη, ης, f.

A Lake	Lacus,ûs, *m.*	2 Λίμνη,ης, f.
A Marsh or Fen	Palus,ūdis, *f.*	1 c. Ἕλος,εος ους, n.
The Sea	Mare,is, *n.*	2 Θάλασσα,ης, f.
The Ocean	Oceănus,ĭ, *m.*	3 Ὠκέανος,άνου, m.
A Wave	Fluctus,ûs, *m.*	5 Κῦμα,ατος, n.
The Tide	Æstus maris, *m.*	5 Κλύδων,ωνος, m.
The Flowing	Æstûs Accessus,ûs, *m.*	2 Πλημμύρα,ας, f.
The Ebb	Æstûs Recessus,ûs, *m.*	5 Ἄμπωτις,ιδος, f.
The Shore	Littus,ŏris, *n.*	3 Αἰγιαλὸς,οῦ, m.
A Ford or Shallow	Vadum,i, *n.*	3 Πόρος,ου, m.
A Deluge	Diluvium,ii, *n.*	3 Κατακλυσμὸς,οῦ, m.
A Whirlpool	Vortex,ĭcis, *m.*	2 Δίνη,ης, f.
A Bay	Sinus,ûs, *m.*	3 Κόλπος,ου, m.
Depth	Profundĭtas,ātis, *f.*	1 c. Βάθος,εος ους, n.
The EARTH	Terra,æ, *f.*	2 Γῆ,γῆς, f.
A Mountain	Mons,tis, *m.*	1 c. Ὄρος,εος ους, n.
A Valley	Vallis,is, *f.*	5 Κοιλὰς,άδος, f.
A Rock	Rupes,is, *f.*	2 Πέτρα,ας, f.
A Plain or Champain	Planities,ei, *f.*	3 Πεδίον,ίου, n.
A Hill	Collis,is, *m.*	3 Βουνὸς,οῦ, m.
A Cliff	Clivus,i, *m.*	3 Κρημνὸς,οῦ, m.
Dirt	Cœnum,ĭ, *n.*	3 Βόρβορος,όρου, m.
Mud	Limus,i, *m.*	5 Ἰλὺς,ύος, f.
Clay	Lutum,i, *n.*	3 Πηλὸς,οῦ, m.
Dust	Pulvis,ĕris, *m.*	2 c. Κόνις,ιος, f.
Sand	Arēna,æ, *f.*	3 Ψάμμος,ου, f.
Gravel	Glarea,æ, *f.*	3 Ἄμαθος,άθου, f.
Quick-sands	Syrtes,ium, *f.*	2 c. Σύρτις,εως, f.
An Island	Insŭla,æ, *f.*	3 Νῆσος,ου, f.

III. Of

III.

Of Stones and Metals.	*De Lapidibus & Metallis.*	Περὶ Λίθων καὶ Μετάλλων.
A Great Stone	Saxum, i, *n.*	2 Πέτρα, ας, f.
A Pebble-Stone	Calcŭlus, i, *m.*	3 Ψῆφος, ου, f.
A Flint-Stone	Silex, ĭcis, *f.*	1 Πυρίτης, ου, m.
A Whet-Stone	Cos, cōtis, *f.*	2 Ἀκόνη, ης, f.
A Touch-Stone	Lydius lapis, *m.*	3 Βάσανος, άνου, f.
A Load-Stone	Magnes, ētis, *m.*	5 Μάγνης, ητος, m.
A Pumice-Stone	Pumex, ĭcis, *d.*	2 c. Κίσηρις, εως, f.
Marble	Marmor, ŏris, *n.*	3 Μάρμαρον, ου, n.
Jet	Gagātes, is, *m.*	1 Γαγάτης, ου, m.
Amber	Succĭnum, i, *n.*	3 Ἤλεκτρον, ου, n.
Alabaster	Alabastrītes, is, *m.*	1 Ἀλαβαστρίτης, ου, m.
An Agate	Achātes, is, *m.*	1 Ἀχάτης, ου, m.
A Jasper	Iaspis, ĭdis, *f.*	5 Ἴασπις, ιδος, f.
Crystal	Crystallus, i, *f.*	3 Κρύσταλλος, ου, m.
Pearl	Margarīta, æ, *f.*	1 Μαργαρίτης, ου, m.
An Emerald	Smaragdus, i, *m.*	3 Σμάραγδος, ου, f.
Brimstone	Sulphur, ŭris, *n.*	3 Θεῖον, ου, n.
Copperas or Vitriol	Vitriŏlum, i, *n.*	3 Χάλκανθος, ου, m.
Ruddle or red Oker	Rubrīca, æ, *f.*	3 Μίλτος, ου, f.
Salt-petre	Nitrum, i, *n.*	3 Νίτρον, ου, n.
Marl	Marga, æ, *f.*	
Chalk	Creta, æ, *f.*	2 Κρητικὴ γῆ, ῆς, f.
Plaster	Gypsum, i, *n.*	3 Γύψος, ου, f.
Pit-coal	Carbo, ōnis, fossilis *m.*	
A Quarry	Lapicidīna, æ, *f.*	2 Λατομία, ας, f.
Glass	Vitrum, i, *n.*	3 Ὕαλος, άλου, f.
A Metal	Metallum, i, *n.*	3 Μέταλλον, άλλου, n.
Minerals	Mineralia, ium, *n.*	

A Mine	Scaptensūla, æ, *f.*	3 Μέταλλον, άλλȣ, n.
Gold	Aurum, i, *n.*	3 Χρυσὸς, ȣ̃, m.
A Mine of Gold	Aurifodīna, æ, *f.*	3 Χρυσωρυχεῖον, είȣ, n.
Silver	Argentum, i, *n.*	3 Ἄργυρος ύρȣ, m.
Brass	Orichalcum, i, *n.*	3 Ὀρείχαλκον, άλκȣ, n.
* *Copper*	Æs, æris, *n.*	3 Χαλκὸς, ȣ̃, m.
Iron	Ferrum, i, *n.*	3 Σίδηρος, ήρȣ, m.
Steel	Chalybs, bis, *m.*	5 Χάλυψ χάλυβος, m.
Lead	Plumbum, i, *n.*	3 Μόλιβδος, ȣ, m.
Tin	Stannum, i, *n.*	3 Κασσίτερος, έρȣ, m.
Quick-silver	Argentum vivum, *n.*	3 Ὑδράργυρος, ύρȣ, m.
Alum	Alūmen, ĭnis, *n.*	2 Στυπτηρία ας, f.
White Lead	Cerussa, æ, *f.*	3 Ψίμμυθος ύθȣ, m.
Saulder	Ferrūmen, ĭnis, *n.*	2 Κόλλα ης, f.
† *Red Lead*	Minium, ii, *n.*	3 Μίλτος, ȣ, f.
Vermilion	Cinnabăris, is, *f.*	2 c. Κιννάβαρι, εως, n.
Verdegrease	Ærūgo, inis, *f.*	3 Ἰὸς, ȣ̃, χαλκȣ̃, m.

* Copper is by modern Writers called *Cuprum*, *fortè q. Æs Cyprium*: and Brass is generally by *English-men* (through mistake) called *Æs*, whereas Brass is not a natural Metal, but a factitious thing of Copper and *Lapis Calaminaris* or *Cadmia*.

† Though modern Writers call Red Lead *Minium*, yet the Antients called it *Sandyx*. The *Minium* of the Antients was twofold, *viz.* Either (1.) the *Cinnabaris* of the Moderns, or (2.) the *Rubrica*.

IV.

Of the Parts & Adjuncts of Plants.	*De Partibus & Appendicib⁹ Plantarū.*	Περὶ τῶν Μερῶν τῶν Φυτῶν.
A *Plant*	Planta, æ, *f.*	3 Φυτὸν, ȣ̃, n.
A Tree	Arbor, ŏris, *f.*	3 Δένδρον, ȣ, n.
A Shrub	Frutex, ĭcis, *m.*	3 Θάμνος, ȣ, m.
An Herb	Herba, æ, *f.*	2 Βοτάνη, ης, f.
A Root	Radix, īcis, *f.*	2 Ῥίζα, ης, f.

The

The Trunk or Body of a tree	Caudex, ĭcis, *m.*	1 c. Στέλεχος, εος ους, n.
A Stalk	Caulis, ĭs, *m.*	3 Καυλὸς, οῦ, m.
Wood	Lignum, i, *n.*	3 Ξύλον, ου, n.
Timber	Materia, æ, *f.*	2 Ὕλη, ης, f.
The Grain of the Wood	Pecten, ĭnis, *m.*	5 Κτήδων ονος, f.
The Pith	Medulla, æ, *f.*	3 Μύελος ου, m. Ἐντεριώνη.
The Sap	Succus, i, *m.*	3 Χυλὸς, οῦ, m.
The Bark	Cortex, ĭcis, *m.*	3 Φλοιὸς οῦ, m.
A Bough or Branch	Ramus, i, *m.*	3 Κλάδος ου m.
A Rod	Virga, æ, *f.*	3 Ῥάβδος ου, f.
A Sprig, Graft or Cyon	Surcŭlus, i, *m.*	5 Κλῶν, νὸς, m.
A Sucker	Stolo, ōnis, *m.*	5 Παραφυὰς άδος, f.
A Bud	Gemma, æ, *f.*	3 Ὀφθαλμὸς, οῦ, m.
A Sprout	Germen, ĭnis, *n.*	3 Βλαστὸς οῦ, m.
A green Leaf	Frons, frondis, *f.*	3 Κλάδος φυλλώδης Πέ-
A Leaf	Folium, ii, *n.*	3 Φύλλον ου, n. (ταλον.
A Blossom or Flower	Flos, floris, *m.*	1 c. Ἄνθος εος ους, n.
Fruit	Fructus, ûs, *m.*	3 Καρπὸς οῦ, m.
A Foot-stalk	Pedicŭlus, i, *m.*	3 Ποδίον ίου, n.
A Prickle	Spina, æ, *f.*	2 Ἄκανθα ης, f.
A Catkin or Palm	Nucamentum, i, *n.*	3 Κύτταρος άρου, m.
The Stone of any Fruit	Ossicŭlum, i, *n.*	2 Πυρήνη ης, f.
A Clasper or Tendrel	Clavicŭla, æ, *n.*	5 Ἕλιξ, ικος, f.
A Cluster	Racēmus, i, *m.*	5 Βότρυς υος, f.
A Husk	Gluma, æ, *f.*	3 Ἔλυτρον ύτρου, n.
A Pod or Cod	Silĭqua, æ, *f.*	3 Κεράτιον ίου, n.
A Shell	Putāmen, ĭnis, *n.*	1 c. Κέλυφος, εος ους, n.
An Awn or Beard	Arista, æ, *f.*	5 Ἀθὴρ έρος, m.
A Kernel	Nucleus, i, *m.*	5 Πυρὴν ῆνος, m.
A Berry	Bacca, æ, *f.*	3 Κόκκος ου m.
A Nut	Nux, nucis, *f.*	3 Κάρυον ύου, n.
An Acorn	Glans, glandis, *f.*	3 Βάλανος άνου, f.
A Cone or Clog	Conus, i, *m.*	3 Κῶνος ου, m.
Brushwood or Bavin	Cremium, ii, *n.*	3 Φρύγιον ίου, n.
A Knot	Nodus, i, *m.*	5 Γόνυ ατος & γουνὸς n.

A Faggot

A Faggot	Faſcis, is, *m.*	2 Δέσμη, ης, f.
Chips	Aſsŭlæ, arum, *f.*	5 Σχίδαξ, ακος, f. σχίδιον, ίου, n. & σκινδαλμὸς, ου, m.

V.

Of Herbs.	*De Herbis.*	Περὶ Βοτανῶν.
1 *Adders-tongue*	Ophiogloſſum, i, *n.*	3 Ὀφιόγλωσσον, ου, n.
2 *Agrimony*	Eupatorium, ii, *n.*	3 Εὐπατώριον ίου, n.
3 *Alexanders*	Hippoſelīnum, i, *n.*	3 Ἱπποσέλινον, ίνου, n.
4 *Aloes*	Aloĕ, es, *f.*	2 Ἀλόη, όης, f.
5 *Anemony or Windflower*	Anemōne, es, *f.*	2 Ἀνεμώνη, ης, f.
6 *Aniſe*	Anīſum, i, *n.*	3 Ἄνισον, ίσου, n.
7 *Archangel*	Lamium, ii, *n.*	2 c. Γαλίοψις, εως, f.
8 *Artichoke*	Cinăra, æ, *f.*	3 Σκόλυμος, ύμου, m.
9 *Aſarabacca*	Asărum, i, *n.*	3 Ἄσαρον, άρου, n.
10 *Aſparagus or Sperage*	Aſparăgus, i, *m.*	3 Ἀσπάραγος, άγου, f.
11 *Avens*	Caryophyllāta, æ, *f.*	
12 *Barley*	Hordeum, ei, *n.*	2 Κριθὴ, ῆς, f.
13 *Baſil*	Ocimum, i, *n.*	3 Ὤκιμον, ίμου, n.
14 *Baulm*	Meliſſophyllum, i, *n.*	3 Μελισσόφυλλον, ου, n.
15 *A Bean*	Faba, æ, *f.*	3 Κύαμος, άμου, m.
16 *Bears-ear*	Auricula urſi, *f.*	
17 *Beet*	Beta, æ, *f.*	3 Τεῦτλον, ου, n.
18 *Betony*	Betonĭca, cæ, *f.*	3 Κέστρον, ου, n.
19 *Bindweed*	Convolvŭlus, i, *m.*	5 Σμίλαξ, ακος, f.
20 *Birthwort*	Ariſtolochia, æ, *f.*	2 Ἀριστολοχία, ας, f.
21 *Bluebottle*	Cyănus, i, *m.*	
22 *Borage*	Borāgo, ĭnis, *f.*	3 Βούγλωσσον, ώσσου, n.
23 *Brooklime*	Anagallis aquatica, *f.*	5 Ἀναγαλλὶς, ίδος, f.
24 *Bryony*	Bryonia, æ, *f.*	2 Βρυωνία, ας, f.

(2.) This is uſually called alſo in Latin *Agrimonia*.
(5.) This in *Engliſh* is commonly, though corruptly, called *Emmony*.
(10.) This alſo is by the vulgar, corruptly called *Sparrow-graſs*.

25 *Buckwheat*

	English	Latin	Greek
25	*Buckwheat*	Fegopyrum, i, *n.*	3 Τραγόπυρον, ύρου, n.
26	*Bugloſs*	Bugloſſum, i, *n.*	3 Βούγλωσσον, ου, n.
27	*A Bur*	Lappa, æ, *f.*	
28	*Burdock*	Perſonāta, æ, *f.*	3 Ἄρκειον, είου, n.
29	*Burnet*	Pimpinella, æ, *f.*	
30	*Cabbage*	Braſsĭca, æ, capitata, *f.*	2 Κράμβη, ης, f.
31	*Calamint*	Calamintha, æ, *f.*	2 Καλαμίνθη, ης, f.
32	*Camomel*	Chamæmēlum, i, *n.*	5 Ἄνθεμις, ιδος, f.
33	*Campions*	Lychnis, idis, *f.*	5 Λυχνὶς, ίδος, f.
34	*Carrot*	Paſtināca, æ, tenuifolia,	3 Σταφυλῖνος, ίνου, m.
35	*Caraways*	Carum, i, *n.*	3 Κάρος, ου, m.
36	*Celandine*	Chelidonium majus.	3 Χελιδόνιον, ίου, n.
37	*Centory*	Centaurium, ii, *n.*	3 Κενταύρειον, ίου, n.
38	*Charlock*	Rapiſtrum, i, *n.*	
39	*Chervil*	Chærephyllum, i, *n.*	
40	*Chickweed*	Alsīne, es, *f.*	2 Ἀλσίνη, ης, f.
41	*Cinquefoil*	Pentaphyllum, i, *n.*	3 Πεντάφυλλον, ου, n.
42	*Clary*	Horminum, i, *n.*	3 Ὅρμινον, ίνου, n.
43	*Cockle*	Pſeudomelāthium, ii, *n.*	3 Ψευδομελάνθιον, ίου, n.
44	*Colts-foot*	Tuſſilāgo, ĭnis, *f.*	3 Βήχιον, ίου, n.
45	*Coriander*	Coriandrum, i, *n.*	3 Κόριον, ίου, n.
46	*Columbine*	Aquilegia, æ, *f.*	
47	*Coſtmary or Alecoſt*	Coſtus, i, hortorum, *f.*	
48	*Cowſlips or Paigles*	Paralyſis, is, *f.*	3 Φλομίσκος, ου, m.
49	*Cranes-bill*	Geranium, ii, *n.*	3 Γεράνιον, ίου, n.
50	*Creſſes*	Naſturtium, ii, *n.*	3 Κάρδαμον, άμου, n.
51	*Crowfoot*	Ranuncŭlus, i, *m.*	3 Βατράχιον, ίου, n.
52	*Colewort*	Braſsĭca, æ, *f.*	2 Κράμβη, ης, f.
53	*Comfrey*	Symphytum, i, *n.*	3 Σύμφυτον, ύτου, n.
54	*A Cucumber*	Cucŭmis, ĕris, *m.*	3 Σίκυον, ου, n.
55	*Cives*	Porrum, i, ſectile *n.*	3 Πράσον, ου, καρτόν, n.

(25) This is known by ſeveral names in ſeveral parts of *England*, as *Brank* in *Eſſex*, &c. *Crap* in *Worceſter-ſhire.*

(28) The uſual name by which this is known in Latin is *Bardana.*

(51) The word *Colewort* was made of the Latin *Caulis*, by which this Plant was ſometimes called, for what reaſon I know not.

56 *Cummin*

56 *Cummin*	Cumīnum, i, *n.*	3 Κύμινον, ίνου, n.
57 *Claver-grass*	Trifolium, ii, majus sativum, *n.*	3 Τρίφυλλον μεῖζον.
58 *Daffodil*	Narcissus, i, *m.*	3 Νάρκισσος, κίσσου, m.
59 *Daisie*	Bellis, idis, *f.*	
60 *Dandelion*	Dens leonis, *m.*	
61 *Darnel*	Lolium, ii, *n.*	2 Αἶρα, ας, f.
62 *Dock*	Lapăthum, i, *n.*	3 Λάπαθον, άθου, n.
63 *Dill*	Anēthum, i, *n.*	3 Ἄνηθον, ήθου, n.
64 *Dragons*	Dracontium, ii, *n.*	3 Δρακόντιον, ίου, n.
65 *Earth-nut or Pig-nut*	Bulbocastănum, i, *n.*	3 Βολβοκαστάνιον, ίου, n.
66 *Elecampane*	Helenium, ii, *n.*	3 Ἑλένιον, ίου, n.
67 *Endive*	Endivia, æ, *f,*	5 Σέρις, ιδος, f.
68 *Eryngo or Sea-Holly*	Eryngium, iì, *n.*	3 Ἠρύγγιον, ίου, n.
69 *Eybright*	Euphrasia, æ, *f.*	2 Εὐφροσύνη, ης, f.
70 *Feverfew*	Matricaria, æ, *f.*	3 Παρθένιον, ίου, n.
71 *Fennel*	Fœnicŭlum, i, *n.*	3 Μάραθρον, άθρου, n.
72 *Fern or Brakes*	Filix, ĭcis, *f.*	5 Πτέρις, ιδος, f.
73 *Flax*	Linum, i, *n.*	3 Λίνον, ου, n.
74 *Flower de luce*	Iris, ĭdis, *f.*	5 Ἶρις, ιδος, f.
75 *Fox-glove*	Digitālis, is, *f.*	
76 *Fumitory*	Fumaria, æ, *f.*	3 Καπνὸς, οῦ, m.
77 *Garlick*	Allium, ii, *n.*	3 Σκόροδον, όδου, n.
78 *Gentian or Felwort*	Gentiāna, æ, *f.*	2 Γεντιανὴ, ῆς, f.
79 *Germander*	Chamædrys, yos, *f.*	5 Χαμαίδρυς, υος, f.
80 *Groundsel*	Senecio, ōnis, *m.*	5 Ἠριγέρων, οντος, m.
81 *Goose-grass or Cleavers*	Aparīne, es, *f.*	2 Ἀπαρίνη, ης, f.
82 *Gourd*	Cucurbĭta, æ, *f.*	2 Κολόκυνθα, ας, f.
83 *Grass*	Gramen, ĭnis, *n.*	2 c. Ἄγρωστις, εως, f.
84 *Gromil or Gromwel*	Lithospermum, i, *n.*	3. Λιθόσπερμον, ου, n.
85 *Harts-tongue*	Lingua cervina, æ, *f.*	5 Φυλλίτις, ιδος, f.
86 *Hawkweed*	Hieracium, ii, *n.*	3 Ἱεράκιον, ίου, n.
87 *Hemlock*	Cicūta, æ, *f.*	3 Κώνειον, είου, n.
88 *Hemp*	Cannăbis, is, *f.*	2 c. Κάνναβις, εως, f.
89 *Hellebore*	Hellebŏrus, i, *f.*	3 Ἑλλέβορος, ου, m.

90 *Henbane*	Hyoscyămus, i, *m.*	3 Ὑοσκύαμος, άμου, m.
91 *Hollyhocks*	Malva, æ, hortensis, *f.*	2 Μαλάχη, ης, f.
92 *Hops*	Lupŭlus, i, *m.*	
93 *Horehound*	Marrubium, ii, *n.*	3 Πράσιον, ίου, n.
94 *Hounds-tongue*	Cynoglossum, i, *n.*	3 Κυνόγλωσσον, ώσσου, n.
95 *House-leek*	Sedum, i, *n.*	3 Ἀείζωον, ώου, n.
96 *Hyssope*	Hyssōpus, i, *f.*	3 Ὕσσωπος, ώπου, f.
97 *A Hyacinth or Jacinth*	Hyacinthus, i, *m.*	3 Ὑάκινθος, κίνθου, m.
98 *A July-flower or Carnation*	Caryophyllus, i, *m.*	3 Καρυόφυλλος, ου, m.
99 *Jack in the hedge*	Alliaria, æ, *f.*	
100 *Kidny or French bean*	Phaseŏlus, i, *m.*	3 Φασίολος, όλου, m.
101 *Larks-spur or heel*	Consolĭda, æ, regalis, *f.*	3 Δελφίνιον, ίου, n.
102 *Lavender*	Lavendŭla, æ, *f.*	
103 *Lavender-Cotton*	Chamæcyparissus, i, *f.*	3 Χαμαικυπάρισσος, ου, f.
104 *Leeks*	Porrum, i, *n.*	3 Πράσον, ου, n.
105 *Lettuce*	Lactūca, æ, *f.*	5 Θρίδαξ, ακος, f.
106 *A Lily*	Lilium, ii, *n.*	3 Κρίνον, ου, n.
107 *Liquorice*	Glycyrrhiza, æ, *f.*	2 Γλυκύῤῥιζα, ης, f.
108 *Liverwort*	Hepatĭca, æ, *f.*	5 Λειχὴν, ῆνος, m.
109 *Lentils*	Lens, tis, *f.*	2 Φακὴ, ῆς, f. & φακός.
110 *Lovage*	Levistĭcum, ci, *n.*	
111 *Lupines*	Lupīnus, i, *m.*	3 Θέρμος, ου, m.
112 *Madder*	Rubia, æ, *f.*	3 Ἐρυθρόδανον, ου, n.
113 *Mallow*	Malva, æ, *f.*	2 Μαλάχη, ης, f.
114 *Mandrake*	Mandragŏra, æ, *f.*	1 Μανδραγόρας, ου, m.
115 *Maiden-hair*	Adianthum, i, *n.*	3 Ἀδίαντον, ου, n.
116 *Marigold*	Calendŭla, æ, *f.*	3. Χρυσάνθεμον, ου, n.
117 *Marjoram*	Majorāna, æ, *f.*	3 Σάμψυχον, ύχου, n.
118 *Herb Mastick*	Marum, i, *n.*	3 Ἀμάρακος, άκου, f.
119 *Maudlin-tansie*	Agerătum, i, *n.*	3 Ἀγήρατον, ου, n.
120 *May-weed*	Cotŭla, æ, *f.*	5 Ἄνθεμις, ιδος, f.
121 *Meadow-sweet*	Ulmaria, æ, *f.*	

(102) This is taken to be a Species of *Stœchas.*
(116) This is thought to be the *Caltha* of the Poets.

122 *Melilot*

122	*Melilot*	Melilōtus, i, *f.*	
123	*Melon*	Melo, ōnis, *m.*	
124	*Mint*	Mentha, æ, *f.*	3 Ἡδύοσμος, όσμου, m.
125	*Moss*	Muſcus, i, *m.*	3 Βρύον, ου, n.
126	*Mouse-ear*	Piloſella, æ, *f.*	5 Μυοσωτὶς, ίδος, f.
127	*Monks-hood*	Napellus, i, *m.*	3 Ἀκόνιτον, ίτου, n.
128	*Mugwort*	Artemiſia, æ, *f.*	2 Ἀρτεμισία, ας, f.
129	*Mullein*	Verbaſcum, i, *n.*	3 Φλόμος, ου, m.
130	*Mustard* (*stool*	Sinapi, is, *n.*	2 c. Σίνηπι, ιος, n.
131	*Mushroom or Toad-*	Fungus, i, *m.*	5 Μύκης, ητος, m.
132	*Millet or Grout*	Milium, ii, *n.*	3 Κέγχρος, ου, m.
133	*Nettle*	Urtīca, æ, *f.*	2 Ἀκαλύφη ης, f. Κνίδη,
134	*Nightshade*	Solānum, i, *n.*	3 Στρύχνος, ου, m. (ης, f.
135	*Oats*	Avēna, æ, *f.*	3 Βρῶμος, ου, m.
136	*Onions*	Cepa, æ, *f.*	3 Κρόμμυον, ου, n.
137	*Orpine*	Telephium, ii, *n.*	3 Τελέφιον, ίου, n.
138	*Orrache*	Atrĭplex, ĭcis, *f.*	2 c. Ἀτράφαξις, εως, f.
139	*Panick*	Panĭcum, ci, *n.*	3 Ἔλυμος, ύμου, m.
140	*Parsly*	Apium, ii, *n.*	3 Σέλινον, ίνου, n.
141	*Parsnip*	Paſtināca, æ, *f.*	3 Ἐλαφόβοσκον, ου, n.
142	*Pease*	Piſum, i, *n.*	3 Πίσον, ου, n.
143	*Pulse* (*grass*	Legūmen, ĭnis, *n.*	3 Ὄσπριον, ίου, n.
144	*Peniroial or Pudding*	Pulegium, ii, *n.*	5 Γλήχων, ωνος, m.
145	*Periwinkle*	Vinca pervinca, æ, *f.*	5 Κληματὶς ίδος, δαφνο-
146	*Pellitory of the wall*	Parietaria, æ, *f.*	2 Ἑλξίνη ης, f. (ειδής, f.
147	*Peiony*	Pæonia, æ, *f.*	2 Παιονία, ας, f.
148	*Pimpernel*	Anagallis, idis, *f.*	5 Ἀναγαλλὶς, ίδος, f.
149	*Plantain*	Plantāgo, ĭnis, *f.*	3 Ἀρνόγλωσσον, ώσσου, n.
150	*Pinks*	Caryophyllus, i, *m.*	3 Καρυόφυλλος, ου, m.
151	*Polypody*	Polypodium, ii, *n.*	3 Πολυπόδιον, ίου, n.
152	*Poppy*	Papāver, ĕris, *n.*	5 Μήκων, ωνος, f.

(150) The Pink and July-flower were not known or at leaſt deſcribed by the Antients, they are by the Moderns called *Caryophylli*, as well from the ſhape of the Flower, together with his Cup reſembling a Clove, as from its ſent.

153 *Prim-*

153	*Primroſe*	Primŭla, æ, veris, *f.*	5 Φλομὶς, ίδος, f.
154	*Purſlane*	Portulāca, æ, *f.*	2 Ἀνδράχνη, ης, f.
155	*Pumpion*	Pepo, ŏnis, *m.*	5 Πέπων, ονος, m.
156	*Radiſh*	Raphănus, i, *m.*	3 Ῥάφανος, άνου, m.
157	*Rampions*	Rapuncŭlus, i, *m.*	
158	*Ramſons*	Allium, ii, urſinum, *n.*	
159	*Reed*	Arundo, ĭnis, *f.*	3 Κάλαμος, άμου, m.
160	*Reſtharrow* or *Camock*	Anōnis, ĭdis, *f.*	5 Ἀνωνὶς, ίδος, f.
161	*Rocket*	Erūca, æ, *f.*	3 Εὔζωμον, ώμου, n.
162	*Rue*	Ruta, æ, *f.*	3 Πήγανον, ου, n.
163	*Rubarb*	Rhabarbarum, i, *n.*	3 Ῥῆον, ου, n.
164	*Ruſh*	Juncus, i, *m.*	3 Σχοῖνος, ου, m.
165	*Rie*	Secāle, is, *n.*	
166	*Rice*	Oryza, æ, *f.*	2 Ὄρυζα, ης, f.
167	*Saffron*	Crocus, i, *m.*	3 Κρόκος, ου, m.
168	*Sage*	Salvia, æ, *f.*	3 Ἐλελίσφακος, ου, m.
169	*Sage of Jeruſalem*	Pulmonaria, æ, maculoſa, *f.*	
170	*Sanicle*	Sanicŭla, æ, *f.*	
171	*Satyrion*	Orchis, idis, *f.*	5 Ὄρχις, ιδος, f.
172	*Savoury*	Satureia, æ, *f.*	2 Θύμβρα, ας, f.
173	*Saxifrage*	Saxifrăga, æ, *f.*	
174	*Scabious*	Scabiōſa, æ, *f.*	2 Ψῶρα, ας, f.
175	*Scallions*	Aſcalonitides, um, *f.*	5 Ἀσκαλωνίτιδες, f.
176	*Scurvy-graſs*	Cochlearia, æ, *f.*	
177	*Skirrets*	Sisărum, i, *n.*	3 Σίσαρον, άρου, n.
178	*Smallage*	Paludapium, ii, *n.*	3 Ἐλειοσέλινον, ου, n.
179	*Sorrel*	Acetōſa, æ, *f.*	5 Ὀξαλὶς, ίδος, f.
180	*Sothern-wood*	Abrotănum, i, *n.*	3 Ἀβρότανον, ου, n.
181	*Sowbread*	Cyclāmen, ĭnis, *n.*	3 Κυκλάμινος, ίνου, f.
182	*Spignel or Meu*	Meum, i, *n.*	3 Μῆον, ου, n.
183	*Spinache*	Spinachia, æ, *f.*	
184	*Spurge*	Tithymalus, i, *m.*	3 Τιθύμαλλος, άλλου, m.
185	*Stock-gilliflower*	Leucoium, ii, *n.*	3 Λευκόϊον, ου, n.
186	*Strawberry*	Fragaria, æ, *f.*	3 Κόμαρον, άρου, n.

187 *Succory*	Cichoreum, i, *n.*	3 Κιχόρειον, είου, n.
188 *A Spunge*	Spongia, æ, *f.*	3 Σπόγγος, ου, m.
189 *Sampire*	Crithmum, i, *n.*	3 Κρίθμον, ου, n.
190 *Stonecrop*	Sedum minus, *n.*	3 Ἀείζωον, ώου, n.
191 *Tansie*	Tanacētum, i, *n.*	
192 *Tares or Vetches*	Vicia, æ, *f.*	3 Βίκιον, ίου, n.
193 *Taragon*	Draco, ōnis, herba, *m.*	5 Τάρχων, οντος, n.
194 *Teasel*	Dipsăcus, i, *m.*	3 Δίψακος, άκου, m.
195 *A Thistle*	Carduus, i, *m.*	3 Σκόλυμος, ύμου, m.
196 *Tormentil*	Tormentilla, æ, *f.*	3 Ἑπτάφυλλον, ου, n.
197 *Trefoil*	Trifolium, ii, *n.*	3 Τρίφυλλον, ου, n.
198 *A Tulip*	Tulĭpa, æ, *f.*	
199 *A Turnep*	Rapum, i, *n.*	2 Γογγύλη, ης, f.
190 *Tyme*	Thymus, i, *m.*	3 Θύμος, ύμου, m.
201 *Tobacco*	Tabacum, ci, *n.*	
202 *Valerian*	Valeriāna, æ, *f.*	3 Νάρδος ἀγρία, f.
203 *Vervain*	Verbēna, æ, *f.*	2 Ἱερὰ βοτάνη, ης, f.
204 *A Violet*	Viŏla, æ, *f.*	3 Ἴον, ου, n.
205 *Wall-flower*	Leucoium, ii, *n.* luteum.	3 Λευκόϊον, ίου, n.
206 *Wheat*	Tritĭcum, ci, *n.*	3 Πυρὸς, οῦ, m.
207 *Sweet Williams*	Armeria, æ, *f.*	
208 *Woad*	Glastum, i, *n.*	5 Ἴσατις, ιδος, f.
209 *Woodruff*	Asperŭla, æ, *f.*	
210 *Wormwood*	Absinthium, ii, *n.*	3 Ἀψίνθιον, ίου, n.
211 *Wrack*	Alga, æ, *f.*	1 c. Φῦκος, εος, ους, n.
212 *Yarrow*	Millefolium, ii, *n.*	3 Χιλιόφυλλος, ου, m.

(201) This is usually called also *Nicotiana*, after the name of him who brought it first over into *France*.

Those Plants for whose Greek names we have left blanks, are partly such as confessedly are not mentioned by any Greek Author now extant, partly such about whose Greek names the modern Herbarists do not agree.

VI. Of

VI.

Of Trees and Shrubs.	*De Arboribus & Fruticibus.*	Περὶ Δένδρων καὶ Θάμνων.
1 AN *Alder-tree*	ALnus, i, *f.*	2 ΚΛῆθρα, ας, f.
2 *An Apple-tree*	Malus, i, *f.*	2 Μηλέα, ας, f.
3 *An Apple*	Pomum, i, *f.*	3 Μῆλον, ου, n.
4 *An Almond-tree*	Amygdălus, i, *f.*	2 Ἀμυγδαλῆ, ῆς, f.
5 *An Almond*	Amygdăla, æ, *f.*	2 Ἀμυγδάλη, ης, f.
6 *An Apricock-tree*	Armeniăca malus.	2 Μηλέα Ἀρμενιακὴ ῆς, f
7 *An Ash*	Fraxĭnus, i, *f.*	2 Μελία ας, f.
8 *A wild Ash*	Ornus, i, *f.*	2 Βουμελία ίας, f.
9 *An Aspen-tree*	Popŭlus, i, *f.* Libyca.	5 Κερκὶς, ίδος, f.
10 *A Barberry-bush*	Oxyacanthus, i, *f.*	2 Ὀξυάκανθα, ας, f.
11 *A Bay-tree*	Laurus, i, *f.*	2 Δάφνη, ης, f.
12 *A Beech-tree*	Fagus, i, *f.*	2 Ὀξύη ης, f.
13 *A Bilberry or Whortle-berry*	Vitis, is, Idæa.	3 Ἄμπελος παρὰ Ἴδης, ἔλου, f.
14 *A Birch-tree*	Betŭla, æ, *f.*	2 Σημύδα, ας, f.
15 *A Box-tree*	Buxus, i, *f.*	3 Πύξος, ου, f.
16 *A Bramble*	Rubus, i, *m.*	3 Βάτος, ου, f.
17 *Broom*	Genista, æ, *f.*	3 Σπάρτιον, τίου, n.
18 *Butchers-broom or Knee-holly*	Ruscus, i, *m.*	2 Ὀξυμυρσίνη, ης, f.
19 *A Cedar-tree*	Cedrus, i, *f.*	3 Κέδρος, ου, f.
20 *A Cherry-tree*	Cerăsus, i, *f.*	3 Κέρασος, άσου, f.
21 *A Chesnut-tree*	Castanea, æ, *f.*	2 Καστάνεια, ας, f.
22 *Capers*	Cappăris, is, *f.*	2 c. Κάππαρις άρεως, f.
23 *A Citron-tree*	Malus Medica.	2 Μηλέα Μηδικὴ ῆς, f.

(10) The Barberry-bush is commonly in Latin called *Berberis*, and is taken by many Herbarists to be the *Oxyacantha* of *Galen*, but not of *Dioscorides*, that being the White-Thorn.

(13) This is usually called in Latin *Vaccinium*, but erroneously as I think.

24 *Cinna*

24	*Cinnamon*	Cinnamōmum, i, *n.*	3 Κιννάμωμον, ου, n.
25	*A Cork-tree*	Suber, ĕris, *n.*	3 Φελλὸς, οῦ, m.
26	*A Cotton-tree*	Gossipium, ii, *n.*	3 Γοσσίπιον, ου, n.
27	*A Cornel-tree or Cornelian Cherry-tree*	Cornus,i, & ûs, *f.*	2 Κρανία, ας, f.
28	*A Cypress-tree*	Cupressus, i, *f.*	3 Κυπάρισσος, ου, f.
29	*Currans*	Ribes, is, *f.*	
30	*Ebony*	Ebĕnus, i, *f.*	3 Ἔβενος, ένου, f.
31	*Elder*	Sambŭcus, i, *f.*	2 Ἀκτὴ ῆς, f.
32	*An Elm*	Ulmus, i, *f.*	2 Πτελέα, ας, f.
33	*A Fig-tree*	Ficus, ûs, *f.*	2 Συκὴ ῆς, f.
34	*A Filberd*	Avellāna nux.	
35	*A Fir-tree*	Abies, ĕtis, *f.*	2 Ἐλάτη, ης, f.
36	*A Fistick Nut-tree or (Pistacho*	Pistacia, æ, *f.*	2 Πιστακία, ας, f.
37	*Furze,Whins or Gorse*	Genista spinosa.	3 Σκόρπιος, ίου, m.
38	*A Gooseberry-bush*	Grossularia, æ, *f.*	3 Ἴσος, ου, f.
39	*Hawthorn orWhitetho.*	Oxyacantha, æ, *f.*	2 Ὀξυάκανθα, ας, f.
40	*A Hazlenut-tree*	Corĭlus, i, *f.*	3 Κάρυον Ποντικὸν, οῦ, n.
41	*Heath*	Erīca, æ, *f.*	2 Ἐρείκη, ης, f.
42	*Holly*	Agrifolium, ii, *n.*	2 Ἀγρία ας, f.
43	*The Holm-Oak*	Ilex, ĭcis, *f.*	3 Πρῖνος, ίνου, f.
44	*An Honey-suckle*	Caprifolium, ii, *n.*	3 Περικλύμενον ένου, n.
45	*The Hornbeam-tree*	Carpinus, i, *f.*	5 Ὄστρυς, υος, f.
46	*Jessamin*	Gelsemīnum, i, *n.*	
47	*Juniper*	Junipĕrus, i, *f.*	3 Ἄρκευθος, εύθου, f.
48	*Ivy*	Hedera, æ, *f.*	3 Κίσσος, ου, f.
49	*Laurel*	Lauroceră̆sus, i, *m.*	

(29) Currants is an equivocal word with us, taken either for the Fruit of a Shrub called in Latin *Ribes*; or a small sort of Grape growing in *Zant*. The name Currant is taken from *Corinthus*, whence its like this Fruit was first brought to us.

(49) The Tree we now commonly call Laurel, is not the Laurel or *Laurus* of the Antients, that being our Bay-tree, but a Plant unknown till of late, bearing an esculent Fruit like a Cherry, and yet an ever-green. This is to be carefully heeded, lest any one be deceived by the confusion of these names.

50 A

50	*A Lemon-tree*	Malus limonia.	
51	*A Lime or Linden-tree*	Tilia, æ, *f.*	1 Φίλυρα, ας, n.
52	*Liquorice*	Glycyrrhiza, æ, *f.*	2 Γλυκυῤῥίζα, ης, f.
53	*A Maple*	Acer, ĕris, *n.*	3 Σφένδαμνος, ου, f.
54	*A Mastick-tree*	Lentiscus, i, *f.*	3 Σχῖνος, ου, f.
55	*A Medlar-tree*	Mespĭlus, i, *f.*	3 Μέσπιλος, ίλου, f.
56	*Misleto*	Viscus, i, *m.*	3 Ἰξός, οῦ m.
57	*A Mulberry-tree*	Morus, i, *f.*	2 Μορέα, ας, f.
58	*A Myrtle-tree*	Myrtus, i, *f.*	2 Μυρσίνη, ης, f.
59	*An Orange-tree*	Malus aurantia.	
60	*An Olive-tree*	Olīva, æ, *f.*	2 Ελαία, ας, f.
61	*An Oak*	Quercus, ûs, *f.*	5 Δρῦς, υός, f.
62	*A Palm-tree*	Palma, æ, *f.*	5 Φοῖνιξ, ικος, f.
63	*A Peach-tree*	Malus Persica.	2 Μηλέα, ας Περσικὴ, f.
64	*A Pear-tree*	Pyrus, i, *f.*	3 Ἄπιος, ίου, f.
65	*A Pine-tree*	Pinus, ûs, *f.*	5 Πίτυς, υος, f.
66	*A Plane-tree*	Platănus, i, *f.*	3 Πλάτανος, άνου, f.
67	*A Plum-tree*	Prunus, i, *f.*	2 Κοκκυμηλέα, ας, f.
68	*A Pomgranat-tree*	Malus Punica.	2 Ῥοιὰ, ᾶς, f.
69	*A Poplar-tree*	Popŭlus, i, *f.*	3 Ἄιγειρος, ου, f.
70	*Privet*	Ligustrum, i, *n.*	3 Κύπρος, ου, f.
71	*A Quince-tree*	Malus Cydonia.	2 Μηλέα Κυδωνία, ας, f.
72	*A Rasp-berry-bush*	Rubus Idæus, *m.*	3 Βάτος Ἰδαία, f.
73	*A Rose*	Rosa, æ, *f.*	3 Ῥόδον, ου, n.
74	*Rose-bay*	Oleander, dri, *m.*	3 Νήριον, ίου, n.
75	*Rosemary*	Ros marinus, i, *m.*	3 Λιβανωτὸς, οῦ, f.
76	*Savin*	Sabīna, æ, *f.*	5 Βράθυς, υος, f.
77	*Saunders*	Santălum, i, *n.*	
78	*Service or Sorb-tree*	Sorbus, i, *f.*	2 Οἴη, ης, f.
79	*Sloe-tree or Black-thorn*	Prunus sylvestris, *f.*	2 Ἀγριοκοκκυμηλέα, ας, f
80	*Strawberry-tree*	Arbŭtus, i, *f.*	3 Κόμαρος, άρου, f.

(50) Whether Lemons and Oranges were known to the Antients is uncertain. Some take them to have been the *Aurea mala Hesperidum*.

	English	Latin	Greek
81	*A Sycamore-tree*	Acer, ĕris, majus, *n.*	3 Σφένδαμνος, ου, f.
82	*A Tamarisk-tree*	Myrīca, a, *f.*	2 Μυρίκη, ης, f.
83	*A Turpentine-tree*	Terebinthus, i, *f.*	3 Τέρμινθος, ου, f.
84	*A Vine*	Vitis, is, *f.*	3 Ἄμπελος, ου, f.
85	*A Walnut-tree*	Juglans, glandis, *f.*	3 Κάρυον Βασιλικὸν οῦ, n.
86	*A Willow-tree*	Salix, ĭcis, *f.*	2 Ἰτέα, ας, f.
87	*A Woodbind or Honey-suckle*	Periclymĕnum, i, *n.*	3 Περικλύμενον, ένου, n.
88	*A Yew-tree*	Taxus, i, *f.*	3 Μίλος, ου, f.
89	*Balm or Balsam*	Balsămum, i, *n.*	3 Βάλσαμον, άμου, n.
90	*A Berry*	Bacca, æ, *f.*	
91	*A Date*	Dactylus, i, *m.*	3 Δάκτυλος, ύλου, f.
92	*An Apple*	Pomum, i, *n.*	3 Μῆλον, ου, n.
93	*A Pear*	Pyrum, i, *n.*	3 Ὄχνη, ης, f.
94	*A Cherry*	Cerăsum, i, *n.*	3 Κεράσιον, ίου, n.
95	*A Plum*	Prunum, i, *n.*	3 Κοκκύμηλον, ήλου, n.
96	*A Fig*	Ficus, ci & cûs, *f.*	3 Σῦκον, ου, n.
97	*An Olive*	Olīvum, i, *n.*	3 Ἔλαιον, αίου, n.
98	*A Nut*	Nux, nŭcis, *f.*	3 Κάρυον, ου, n.
99	*A Nut-shell*	Putāmen, ĭnis, *n.*	1 c. Κέλυφος, εος ους, n.
100	*A Kernel*	Nucleus, ei, *m.*	5 Πυρὴν, ῆνος, m.
101	*A Walnut*	Juglans, dis, *f.*	3 Κάρυον Βασιλικὸν ου, n.
102	*A Chesnut*	Castanea, æ, *f.*	3 Βάλανος Διὸς, f.
103	*A Nut-cracker*	Nucifrangibulum, i, *n.*	
104	*A Quince*	Cydonium, ii, *n.*	3 Κυδώνιον, ίου, n.
105	*An Orange*	Aurantium, ii, *n.*	
106	*A Citron*	Medicum, i, *n.*	3 Μηδικὸν, οῦ, n.
107	*A Lemon*	Limonium, ii, *n.*	
108	*A Warden*	Volēmum, i, *n.*	
109	*An Apricock*	Malum Armeniăcum, i,	3 Μῆλον Ἀρμενιακόν.
110	*A Peach*	Malum Persicum, i, *n.*	3 Περσικὸν, οῦ, n.

(81) That we vulgarly but corruptly call the Sycomore-tree is not the *Sycomorus* of the Antients or the Tree so called in Scripture, but a sort of Maple. I suppose it was first so mis-named because the Leaf resembles a Fig-Leaf.

111 A

111	*A Strawberry*	Fragum, i, *n.*	3 Κόμαρον, άρυ, n.
112	*A Blackberry*	Morum rubi, *n.*	
113	*A Mulberry*	Morum, i, *n.*	3 Μόρον, υ, n.
114	*A Goosebery*	Grossŭla, æ, *f.*	
115	*Currans*	Uvæ Corinthiacæ, *f.*	
116	*Sugar*	Sacchărum, i, *n.*	3 Σάκχαρον, άρυ, n.
117	*Pepper*	Piper, ĕris, *n.*	2 c. Πέπερι, έρεως, n.
118	*Ginger*	Zingĭber, ĕris, *n.*	2 c. Ζιγγίβερις, έρεως, n.
119	*Nutmeg*	Nux, cis, moschata, æ, *f.*	3 Κάρυον ἀρωματικὸν, n.
120	*Mace*	Macis, is, *f.*	
121	*Frankincense*	Thus, thūris, *n.*	3 Λίβανος άνυ, m.
122	*Myrrh*	Myrrha, æ, *f.*	2 Μύῤῥα, ας, f.
123	*Rosin*	Resīna, æ, *f.*	2 Ῥητίνη, ης, f.
124	*Turpentine*	Terebinthĭna, æ, *f.*	2 Τερμινθίνη, ης, f.
125	*Pitch*	Pix, picis, *f.*	2 Πίσσα, ης, f.
126	*A Clove*	Caryophyllus aromaticus, i, *m.*	3 Καρυόφυλλος, υ, n.
127	*An Acorn*	Glans, glandis, *f.*	3 Βάλανος, άνυ, f.
128	*A Gall*	Galla, æ, *f.*	5 Κηκὶς, ίδος, f.

(125) The word *Pix* is a name common to Pitch and Tar: Tar being called *Pix liquida*, and Pitch *Pix arida* or *navalis*, and in Greek Πισσίματα.

VII.

Of the proper Parts and Adjuncts of Animals.	*De Partibus propriis & Adjunctis Animalium.*	Περὶ τῶν Μερῶν τῶν Ζώων.
I. Of Fishes.	I. *Piscium.*	Ἰχθύων.
A Scale	Squama, æ, *f.*	5 Λεπὶς ίδος, f.
A Shell	Testa, æ, *f.*	3 Ὄστρακον άκυ, n.
A Shelfish	Concha, æ, *f.*	2 Κόγχη, ης, f.
The Gills of a Fish	Branchiæ, arum, *f.*	3 Βράγχια, ίων, n.
The Fins	Pinnæ, arum, *f.*	3 Πτερύγια, ίων, n.

The

The Milt	Lac piſcium, *n.*	3 Θορὸς, οῦ, m.
The Row or Spawn	Ova piſcium, *n.* Lactes.	3 Ὠὰ ἰχθύων.
A Fiſherman	Piſcātor, ōris, *m.*	3 c. Ἁλιεὺς, έως, m.
A Net	Rete, is, *n.*	3 Δίκτυον, ύου, n.
An Angle-rod	Arundo, ĭnis, *f.*	3 Κάλαμος, άμου, m.
A Line	Linea, æ, piſcatoria, *f.*	2 Ὁρμιὰ, ᾶς, f.
A Hook	Hamus, i, *m.*	3 Ἄγκιστρον, ίστρου, n.
A Bait	Eſca, æ, *f.*	5 Δέλεαρ, έατος, n.
A Plummet	Bolis, ĭdis, *f.*	5 Βολὶς, ίδος, f.
Fiſhing	Piſcatūra, æ, *f.*	2 Ἁλιεία, ας, f.
A Fiſhmonger	Ichthyopōla, æ, *m.*	1 Ἰχθυοπώλης, ου, m.
Salt-Fiſh	Salſamentum, i, *n.*	1 c. Τάριχος, εος, ους, n.
II. Of Birds.	**II.** ***Avium.***	**Ὀρνίθων.**
A Birds Bill or Beak	Roſtrum, i, *n.*	1 c. Ῥύγχος, εος, ους, n.
A Comb or Creſt	Criſta, æ, *f.*	3 Λόφος, ου, m.
A Gill or Wattle	Palea, æ, *f.*	3 Κάλλαιον, αίου, n.
A Wing	Ala, æ, *f.*	3 Πτερὸν, οῦ, n.
The Craw or Crop	Ingluvies, ei, *f.*	3 Πρόλοβος, ου, m.
The Rump	Orrhopygium, ii, *n.*	3 Ὀρροπύγιον, ίου, n.
A Claw or Talon	Unguis, is, *m.*	5 Ὄνυξ, υχος, m.
A Feather	Pluma, æ, *f.*	3 Πτίλον, ου, n.
A hard Feather	Penna, æ, *f.*	3 Πτέρον, ου, n.
A Quill	Calămus, i, *m.*	3 Κάλαμος, άμου, m.
Down	Lanūgo, ĭnis, *f.*	3 Χνόος, όου, m.
A Spur	Calcar, āris, *n.*	3 Πλῆκτρον, ου, n.
A Neſt	Nidus, i, *m.*	2 Νεοττία, ας, f.
An Egg	Ovum, i, *n.*	3 Ὠὸν, οῦ, n.
The Yolk	Vitellus, i, *m.*	3 Λέκιθος, ίθου, f.
The White	Albūmen, ĭnis, *n.*	5 Λεύκωμα, ατος, f.
An Egg-ſhell	Teſta ovi.	1 c. Κέλυφος, εος, ους, n.
A Fowler	Auceps, cŭpis, *m.*	1 Ὀρνιθοθήρας, ου, m.
Birdlime	Viſcus, ci, *m.*	3 Ἰξὸς, οῦ, m.
A Cage or Aviary	Aviarium, ii, *n.*	5 Ὀρνιθὼν, ῶνος, m.

III. Of

III. Of Beasts.	III. *Quadrupedum.*	Τετραπόδων.
Cattel	Pecus, ŏris, *n.*	5 Θρέμμα, ατος, n.
An Herd	Armentum, i, *n.*	2 Ἀγέλη, ης, f.
A labouring Beast	Jumentum, i, *n.*	1 c. Κτῆνος, εος, ους, n.
A wild Beast	Fera, æ, *f.*	5 Θὴρ, ρός, m.
A Hide	Corium, ii, *n.*	2 Βύρσα, ης, f.
A Horn	Cornu, *indecl. n.*	5 c. Κέρας, ατος, αος, ως, n.
A Hoof	Ungŭla, æ, *f.*	2 Ὁπλὴ, ῆς, f.
Hair	Pilus, i, *m.*	5 Θρὶξ, τριχὸς, f.
A Bristle	Seta, æ, *f.*	2 Χαίτη, ης, f.
A Mane	Juba, æ, *f.*	2 Χαίτη, ης, f.
Wool	Lana, æ, *f.*	3 Ἔριον, ίου, n.
A Fleece	Vellus, ĕris, *n.*	3 Μαλλὸς, οῦ, m.
A Tail	Cauda, æ, *f.*	2 Οὐρὰ, ᾶς, f.
A Trunk	Proboſcis, ĭdis, *f.*	5 Προβοσκὶς, ίδος, f.

VIII.

Of Four-footed Beasts.	*De Quadrupedibus.*	Περὶ Τετραπόδων.
1 AN *Ape*	SImia, æ, *f.*	3 Πίθηκος, ου, m.
2 *An Aſs*	Asĭnus, i, *m.*	3 Ὄνος, ου, m.
3 *A wild Aſs*	Onăger, ri, *m.*	3 Ὄναγρος, άγρου, m.
4 *A Baboon*	Papio, ōnis, *m.*	
5 *A Badger, Brock or Gray*	Melis, is, *f.* Taxus, i, *m.*	
6 *A Bat*	Veſpertilio, ōnis, *m.*	5 Νυκτερὶς, ίδος, f.
7 *A Bear*	Urſus, i, *m.*	3 Ἄρκτος, ου, com.

(4) This Animal is by ſome taken to be the *Cynocephalus* of the Antients, but erroneouſly.

(6) A Bat though it flies, hath no affinity to Birds, not ſo much as a flying Serpent: and though it be not properly a Quadruped, yet hath it Claws in the Wings, which anſwer to Fore-Legs.

8 *A Beaver*	Fiber, bri, *m.*	5 Κάστωρ, ορος, m.
9 *A Buffle*	Bubălus, i, *m.*	
10 *A Camel*	Camēlus, i, *m.*	3 Κάμηλος, ήλου, com.
11 *A Cat*	Felis, is, *f.*	3 Αἴλουρος, ούρου, m.
12 *A Chameleon*	Chamæleo, ontis, *m.*	5 Χαμαιλέων οντος, m.
13 *A Coney*	Cunicŭlus, i, *m.*	5 Δασύπους, οδος, m.
14 *A Crocodile*	Crocodīlus, i, *m.*	3 Κροκόδειλος, ου, m.
15 *A Deer*	Dama, æ, *com.*	5 Πρὸξ, κός, f.
16 *A Fawn*	Hinnŭlus, i, *m.*	3 Νεβρὸς, οῦ, m.
17 *A Hart or Stag*	Cervus, i, *m.* }	3 Ἔλαφος, άφου, com.
18 *A Hind*	Cerva, æ, *f.* }	
19 *A Pricket*	Subŭlo, ōnis, *m.*	
20 *A Roe*	Caprea, æ, *f.*	5 Δορκὰς, άδος, f.
21 *A Dog or Bitch*	Canis, is, *com.*	5 Κυὼν, νός, com.
22 *A Whelp*	Catŭlus, i, *m.*	3 Σκύμνος, ου, m.
23 *A Mastive*	Molossus, si, *m.*	3 Μολοσσὸς, οῦ, m.
24 *A Spaniel*	Hispaniŏlus, i, *m.*	
25 *A Mungrel*	Hybris, ĭdis, *f.*	1 c. Ἑτερογενὴς, έος, ους.
26 *A Dogs Collar*	Millum, i, *n.*	
27 *Barking*	Latrātus, ûs, *m.*	3 Ὑλαγμὸς, οῦ, m.
28 *A Muzzle*	Fiscella, æ, *f.*	3 Κημὸς, οῦ, m.
29 *A Dragon*	Draco, ōnis, *m.*	5 Δράκων, οντος, m.
30 *A Dromedary*	Dromedarius, ii, *m.*	5 Δρομὰς, άδος, f.
31 *An Elephant*	Elephas, antis, *m.*	5 Ἐλέφας, αντος, m.
32 *An Elk*	Alce, is, *f.*	
33 *A Ferret*	Viverra, æ, *f.*	5 Ἰκτὶς, ίδος, f.
34 *A Fox*	Vulpes, is, *f.*	5 Ἀλώπηξ, εκος, f.
35 *A Frog*	Rana, æ, *f.*	3 Βάτραχος, άχου, m.
36 *A He-Goat*	Hircus, i, *m.*	3 Τράγος, ου, m.
37 *A She-Goat*	Capra, æ, *f.*	5 Αἲξ, γός, f.
38 *A young Goat or Kid*	Hœdus, i, *m.*	3 Ἔριφος, ίφου, m.
39 *A Goat-herd*	Caprarius, ii, *m.*	3 Αἰπόλος, ου, m.

(9) The name Buffle is so near in sound to *Bubalus*, that the one must needs, I think, be derived from the other; howbeit I am not ignorant that some Naturalists will not allow them to be the same.

40 *A*

40	*A Hare*	Lepus, ŏris, *m.*	3 Λαγωὸς, ῦ, m.
41	*A Leveret or yong Hare*	Lepuſcŭlus, i, *m.*	2 Λαγίδιον, ίου, n.
42	*A Hedghog*	Echīnus, i, *m.*	3 Ἐχῖνος, ου, m.
43	*A Hog*	Porcus, i, *m.*	3 Χοῖρος, ου, m.
44	*A Hog-ſty*	Suile, is, *n.*	3 Συφεὸς, ῦ, m.
45	*An Hog-herd*	Subulcus, ci, *m.*	3 Χοιροβοσκὸς, ῦ, m.
46	*A Boar*	Aper, ri, *m.*	3 Κάπρος, ου, m.
47	*A Sow*	Sus, ſuis, *com.*	5 Ὗς, ὑός, com.
48	*A Pig*	Porcellus, i, *m.*	3 Χοιρίδιον, ίου, n.
49	*A Horſe*	Equus, i, *m.* }	3 Ἵππος, ου, com.
50	*A Mare*	Equa, æ, *f.* }	
51	*A Gelding*	Spado, ōnis, *m.*	5 Σπάδων, ωνος, m.
52	*A Race-horſe*	Celes, ĭtis, *m.*	5 Κέλης, ητος, m.
53	*A Nag*	Mannus, i, *m.*	3 Ἱππάριον, ίου, n.
54	*An ambling Nag*	Equus tolutarius aut gradarius.	
55	*A War-Horſe*	Equus bellatorius.	3 Ἵππος πολεμικὸς, m.
56	*A Trotter*	Succuſſarius, ii, *m.*	
57	*A Hackney-Horſe*	Equus meritorius, *m.*	3 Ἵππος μισθωτικός.
58	*A Pack-Horſe*	Equus clitellarius, *m.*	3 Ἵππος ἐπισακτικός.
59	*Neighing*	Hinnītus, ûs, *m.*	3 Χρεμετισμὸς, ῦ, m.
60	*An Horſe-keeper*	Equĭſo, ōnis, *m.*	3 Ἱπποκόμος, ου, m.
61	*An Halter*	Capiſtrum, i, *n.*	3 Κημὸς, ῦ, m.
62	*A Jack-all*	Lupus aureus, *m.*	
63	*A Leopard*	Pardus, i, *m.*	2 c. Πάρδαλις, εως, f.
64	*A Lion*	Leo, ōnis, *m.*	5 Λέων, οντος, m.
65	*A Lioneſs*	Leæna, æ, *f.*	2 Λέαινα, ης, f.
66	*A Lizard*	Lacertus, i, *m.*	3 Σαῦρος, ου, m.
67	*A Mole or want*	Talpa, æ, *d.*	5 Ἀσπάλαξ, κος, m.
68	*A Mole-Hill*	Grumŭlus, i, *m.*	3 Θρομβίον, ου, n.
69	*A Monkey*	Cercopithēcus, i, *m.*	3 Κερκοπίθηκος, ήκου, m.
70	*A Mouſe*	Mus, ūris, *m.*	5 Μῦς, μυός, m.
71	*A Dormouſe*	Glis, gliris, *m.*	3 Ἐλειὸς, ῦ, m.

(71) That little Beaſt which with us is uſually called a Dormouſe is not *Glis*, but *Mus avellanarum* of Naturaliſts. The *Glis* is unknown to me, and therefore becauſe it is vulgarly taken for our Dormouſe, I have let it ſo ſtand.

72 *A Mouſe-trap*	Muſcipŭla, æ, *f.*	2 Μυάγρα, γρας, f.
73 *A Rat or Mouſ-catcher*	Muricīda, æ, *m.*	3 Μυοφόνος, ου, com.
74 *A Mule*	Mulus, i, *m.*	3 Ἡμίονος, ου, com.
75 *An Ox*	Bos, bŏvis, *com.*	5 Βοῦς, βοός, com.
76 *A Bull*	Taurus, i, *m.*	3 Ταῦρος, ου, m.
77 *A Cow*	Vacca, æ, *f.*	2 c. Δάμαλις άλεως, f.
78 *A Heifer*	Juvenca, æ, *f.*	2 c. Πόρταλις άλεως, f.
79 *A Bullock*	Juvencus, i, *m.*	5 Πόρταξ, κος, m.
80 *A Calf*	Vitŭlus, i, *m.*	3 Μόσχος, ου, m.
81 *An Udder*	Mamma, æ, *f.*	3 Μαστὸς οῦ, m.
82 *A Teat*	Papilla, æ, *f.*	2 Θηλὴ, ῆς, f.
83 *A Porcupine*	Hyſtrix, ĭcis, *f.*	5 Ὕστριξ, χος, f.
84 *A Sheep*	Ovis, is, *f.*	3 Πρόβατον, άτου, n.
85 *A Ram*	Aries, ĕtis, *m.*	3 Κριός, οῦ, m.
86 *A Weather*	Vervex, ēcis, *m.*	3 Κριὸς ἐκτετμημένος.
87 *A Lamb*	Agnus, i, *m.*	3 Ἀμνὸς, οῦ, m.
88 *A Sheepfold*	Ovīle, is, *n.*	2 Ποιμνὴ, ῆς, f.
89 *A Shepherd*	Paſtor, ōris, *m.*	5 Ποιμὴν, ένος, m.
90 *A Sheephook*	Pedum, i, *n.*	5 Καλαῦροψ, οπος, f.
91 *A Flock*	Grex, grĕgis, *m.*	3 Ποίμνιον, ίου, n.
92 *Wool*	Lana, æ, *f.*	3 Ἔριον, ίου, n.
93 *A Fleece*	Vellus, ĕris, *n.*	3 Πόκος, ου, m.
94 *A Squirrel*	Sciūrus, i, *m.*	3 Σκίουρος, ούρου, m.
95 *A Tiger*	Tigris, is, vel ĭdis, *f.*	5 Τίγρις, ιδος, f.
96 *A Toad*	Bufo, ōnis, *m.*	2 Φρύνη, ης, f.
97 *A Panther*	Panthēra, æ, *f.*	5 Πάνθηρ, ηρος, m.
98 *A Weaſel*	Muſtēla, æ, *f.*	2 Γαλῆ, ῆς, f.
99 *A Martin or Sable*	Martes, is, *f.*	
100 *A Polcat*	Putorius, ii, *m.*	
101 *An Ounce*	Lynx, cis, *com.*	5 Λύγξ, γκός, f.
102 *An Otter*	Lutra, æ, *f.*	2 c. Ἔνυδρις, εως, f.
103 *A Wolf*	Lupus, i, *m.*	3 Λύκος, ου, m.
104 *A Seal or Sea-calf*	Phoca, æ, *f.*	2 Φώκη, ης, f.
105 *A Tortoiſe*	Teſtūdo, ĭnis, *f.*	2 Χελώνη, ης, f.
106 *A Tadpole*	Gyrīnus, i, *m.*	3 Γυρῖνος, ου, m.

107 *A Salamander*	Salamandra, æ, *f.*	2 Σαλαμάνδρα, ας, f.
108 *An Evet or Newt*	Salamandra aquatica.	
109 *A Rat*	Sorex, ĭcis, *m.*	5 Μῦς μείζων, m.
110 *A Serpent*	Serpens, tis, *m.*	2 c. Ὄφις, εως, m.
111 *A Snake*	Anguis, is, *com.*	
112 *A Water-Snake*	Hydra, æ, *f.*	2 Ὕδρα, ας, f.
113 *A Slow-worm*	Cæcilia, æ, *f.*	3 Τυφλῖνος, ου, m.
114 *An Adder or Viper*	Vipera, æ, *f.*	2 Ἔχιδνα, ης, f.

(109) Some make *Sorex* to signifie the Field or Shrew-Mouse, the Rat they call in Latin *Rattus* or *Mus domesticus major.*

(111) That Serpent which we in English call a Snake, is called by Writers of the History of Animals in Latin *Natrix torquata.*

(114) The English Adder is the same with the Viper. I am sure it is the same that the Apothecaries take and use for the Viper, both in *France* and *Italy*, and what modern Naturalists esteem and describe for the Viper; and a sufficient Argument that it is the Viper of the Antients is, that it bringeth forth living young.

IX.

Of Birds.	*De Avibus.*	Περὶ Ὀρνίθων.
1 *A Bittour or Bittern*	Ardea stellaris, *f.*	1 Ἀστερίας, ου, m.
2 *A Blackbird*	Merŭla, æ, *f.*	3 Κόσσυφος, ου, m.
3 *A Brambling*	Montifringilla, æ, *f.*	1 Ὀρεοσπίζης, ου, m.
4 *A Bulfinch*	Rubicilla, æ, *f.*	1 Πυῤῥούλας, ου, m.

(1) This Bird is supposed to be the *Taurus* of *Pliny*. The Author of *Philomela* calls it *Butio*, but his Authority is of no great weight. Some of the Moderns make this Bird the *Onocrotalus* of the Antients, because of the noise he makes: But the *Onocrotalus* is now acknowledged to be a far different Fowl, *viz.* that we call in English the Pelecan, and is by Naturalists also so taken to be.

(4) This Bird is also called in English an *Alp* or *Nope.*

(4) The word *Rubicilla* is no antient Latin word, but imposed by *Gaza*, who so rendred the Greek Πυῤῥούλας in *Aristotle.*

5 A

5 *A Bunting*	Emberiza, æ, *f.* alba.	
6 *A Bustard*	Tarda, æ, *f.*	5 Ὠτὶς, ίδος, f.
7 *A Buzzard*	Buteo, ōnis, *m.*	1 Τριόρχης, ου, m.
8 *A Capon*	Capo, ōnis, *m.*	
9 *A Chaffinch*	Fringilla, æ, *f.*	2 Σπίζα, ης, f.
10 *A Chough*	Coracias, æ, *m.*	1 Κορακίας, ου, m.
11 *A Coot*	Fulĭca, æ, *f.*	
12 *A Cormorant*	Corvus, i, aquaticus, *m.*	5 Κόραξ, ακος, m.
13 *A Cock*	Gallus, i, *m.*	5 Ἀλέκτωρ, ορος, f.
14 *A Cocks Comb*	Crista, æ, *f.*	3 Ἀλεκτορόλοφος, ου, m.
15 *A Crane*	Grus, gruis, *com.*	3 Γέρανος, ου, m.
16 *A Cuckow*	Cucūlus, i, *m.*	5 Κόκκυξ, υγος, m.
17 *A Curlew*	Arquāta, æ, *f.*	3 Νουμήνιος, ου, m.
18 *A Didopper*	Colymbus, i, *m.*	3 Κόλυμβος, ου, m.
19 *A Dottrel*	Morinellus, i, *m.*	
20 *A Dove or Pigeon*	Columba, æ, *f.*	2 Περιστερὰ, ας, f.
21 *A Turtle-Dove*	Turtur, ŭris, *m.*	5 Τρυγὼν, όνος, f.
22 *A Ring-Dove*	Palumbus, i, *m.*	2 Φάττα, ης, f.
23 *A Stock-Dove*	Vināgo, ĭnis, *f.*	5 Οἰνὰς, άδος, f.
24 *A Dove-House*	Columbarium, ii, *n.*	5 Περιστερεὼν, ῶνος, m.

(5) *Emberiza* is no antient Latin word, but made by *Gesner* of the Dutch **Emmeritz**. This Bird is by some named *Rubetra*, and by others *Calandra*.

(10) This Bird is also called *Pyrrhocorax*, from the redness of the Bill and Legs; though I am not ignorant that *Aldrovandus* makes *Pyrrhocorax* and *Coracias* two distinct Birds.

(16) Note, that the *Penult.* in *Cuculus* is long, notwithstanding that the Author of *Philomela* makes it short.

(17) The word *Arquata* is no antient Latin name, but imposed on this Bird by *Gesner*, from the bending Bill resembling a Bow.

(18) This Bird is also called a Doucker and a Dobchick: The word *Colymbus* is by later Naturalists appropriated to this kind, and *Mergus* used for another sort of Divers more like to Ducks, having toothed Bills; though indeed *Mergus* be a general name common to both sorts.

(19) *Morinellus* is also no antient Latin word, but a new name imposed upon this Bird by Dr. *Caius*, *à Morinis populis, quibus nobiscum communis est*, and ἀπὸ τῆς μωρότητος.

25	*A Duck*	Anas, ătis, *com.*	2 Νῆττα, ης, f.
26	*An Eagle*	Aquĭla, æ, *f.*	3 Ἀετὸς, ȣ̃, m.
27	*A Falcon*	Falco, ōnis, *m.*	
28	*A Fieldfare*	Turdus, i, pilaris, *m.*	5 Τειχὰς, άδος, f.
29	*A Goldfinch*	Carduĕlis, is, *f.*	5 Ἀκανθὶς, ίδος, f.
30	*A Goose*	Anſer, ĕris, *m.*	5 Χὴν, ηνὸς, com.
31	*A Greenfinch*	Chloris, ĭdis, *f.*	5 Χλωρὶς, ίδος, f.
32	*A Griffon*	Gryps, gryphis, *m.*	5 Γρὺψ, πός, m.
33	*A Gull or Sea-mew*	Larus, i, *m.*	3 Λάρος, ȣ, m.
34	*A Hawk*	Accipĭter, tris, *m.*	5 Ἱέραξ, ακος, m.
35	*A Heathcock*	Tetrao, ōnis, *m.*	5 Τέτραξ, γος, f.
36	*A Hedg-ſparrow*	Currūca, æ, *f.*	5 Ὑπολαῒς, ίδος, f.
37	*An Hen*	Gallīna, æ, *f.*	5 Ὄρνις, ιθος, f.
38	*An Heron*	Ardea, æ, *f.*	3 Ἐρωδιὸς, ȣ̃, m.
39	*A Hobby*	Subbuteo, ōnis, *m.*	1 Ὑποτριόρχης, ȣ, m.
40	*A Hoop or Hoopo*	Upŭpa, æ, *f.*	5 Ἔποψ, πος, m.
41	*A Jack-daw*	Monedŭla, æ, *f.*	3 Κολοιὸς, ȣ̃, m.
42	*A Jay*	Pica, æ, glandaria, *f.*	1 Μελαγκοκράνης, ȣ, m.
43	*A Keſtrel*	Tinnuncŭlus, i, *m.*	5 Κεγχρὶς, ίδος, f.
44	*A Kingsfiſher*	Iſpĭda, æ, *f.*	
45	*A Kite or Glead*	Milvus, i, *m.*	3 Ἰκτῖνος, ȣ, m.
46	*A Lapwing*	Vannellus, i, *m.*	5 Αἴξ, γός, f.
47	*A Lark*	Alauda, æ, *f.*	3 Κορύδαλος, ȣ, m.

(33) This Bird is called alſo a Sea-cob, and in Latin *Gavia.*

(35) This is alſo called a *Grouſe*, one ſort heerof is taken to be the ſo much commended *Attagen* of the Antients, *viz.* that called the *red Game.*

(44) Our Kingfiſher is vulgarly taken for the *Halcyon* of the Antients, but the notes of the *Halcyon* do not all agree to it, and therefore modern Writers have impoſed the name *Iſpida* upon it, *à ſono vocis.*

(46) *Vannellus* is a new-made name of the French *Vanneau.* This Bird by a great miſtake hath been generally taken to be the *Upupa* of the Antients, which is now by all acknowledged to be the *Hoopo.*

(47) This Bird is alſo called in Latin *Caſſita* and *Galerita,* becauſe one ſort of it hath a tuft or creſt on the head.

48 *A*

	English	Latin	Greek
48	*A Linnet*	Linaria, æ, *f.*	
49	*A Martin*	Hirundo, ĭnis, *f.* agrestis.	3 Κύψελος, ου, m.
50	*A Merlin*	Æsălon, ōnis, *m.*	5 Ἀισάλων, νος, m.
51	*A More-hen*	Gallinŭla, æ, *f.*	
52	*A Nightingale*	Luscinia, æ, *f.*	5 Ἀηδὼν, όνος, f.
53	*A Night-Raven*	Nycticŏrax, ācis, *m.*	5 Νυκτικόραξ, κος, m.
54	*A Nuthatch*	Sitta, æ, *f.*	2 Σίττη, ης, f.
55	*An Ostrich*	Struthiocamēlus, i, *m.*	3 Στρουθοκάμηλος, ου, m.
56	*An Owl*	Noctua, æ, *f.*	5 Γλαὺξ, κός, f.
57	*A Screech-Owl*	Strix, igis, *f.*	5 Στρὶξ, γός, f.
58	*A Parrot*	Psittăcus, i, *m.*	2 Ψιττάκη, ης, f.
59	*A Partridge*	Perdix, īcis, *com.*	5 Πέρδιξ, κος, com.
60	*A Peacock*	Pavo, ōnis, *m.*	4 Ταὼς, ῶ, m.
61	*A Pelecan*	Pelecānus, i, *m.*	5 Πελεκὰν, ᾶνος, m.
62	*A Pheasant*	Phasiānus, i, *m.*	3 Φασιανὸς, οῦ, m.
63	*A Phœnix*	Phœnix, īcis, *m.*	5 Φοῖνιξ, κος, m.
64	*A Pie or Magpie*	Pica, æ, *f.*	2 Κίττα, ης, f.
65	*A Plover*	Pluviālis, is, *f.*	3 Πάρδαλος, άλου, m.
66	*A Quail*	Coturnix, īcis, *f.*	5 Ὄρτυξ, γος, m.
67	*A Raven*	Corvus, i, *m.*	5 Κόραξ, κος, m.
68	*A Rail*	Ortygomētra, æ, *f.*	2 Ὀρτυγομήτρα, ας, f.
69	*A Redstart*	Ruticilla, æ, *f.*	3 Φοινίκουρος, ου, m.
70	*Robin red-brest*	Rubecŭla, æ, *f.*	3 Ἐρίθακος, ου, m.
71	*A Rook*	Cornix, īcis, *f.* frugilega	2 Κορώνη σπερμολόγος, f.
72	*A Siskin*	Spinus, i, *m.*	3 Σπίνος, ου, m.
73	*A Snipe or Snite*	Gallināgo, ĭnis, minor.	5 Σκολόπαξ, κος, ἐλάττων, m.
74	*A Sparrow*	Passer, ĕris, *m.*	3 Στρουθὸς, οῦ, m.
75	*A Starling or Stare*	Sturnus, i, *m.*	5 Ψὰρ, ρός, m.
76	*A Stork*	Ciconia, æ, *f.*	3 Πελαργὸς, οῦ, m.
77	*A Swallow*	Hirundo, ĭnis, *f.*	5 Χελιδὼν, όνος, f.
78	*A Swan*	Cygnus, i, *m.*	3 Κύκνος, ου, m.

(48) This is also no antient Latin word, but a newly imposed name, as is also *Gallinula* for a More-hen, *Pluvialis* for a Plover, *Ruticilla* for a Redstart, *Rubecula* for a Robin red-breast.

80 *A*

81 *A Teal*	Querquedŭla, æ, *f.*	5 Φασκὰς ἀδος, m.
82 *A Thrush*	Turdus, i, *m.*	2 Κίχλη, ης, f.
83 *A Titmouse*	Parus, i, *m.*	3 Ἀιγίθαλος, ου, m.
84 *A Turkey*	Gallopāvo, ōnis, *m.*	5 Μελεαγρὶς, ίδος, f.
85 *A Vultur*	Vultur, ŭris, *m.*	5 Γύψ, πός, m.
86 *A Wagtail*	Motacilla, æ, *f.*	5 Σεισοπυγὶς, ίδος, f.
87 *A Widgeon*	Penelŏpe, es, *f.*	5 Πηνέλοψ, οπος, f.
88 *A Woodcock*	Scolŏpax, ācis, *m.*	5 Σκολόπαξ, κος, m.
89 *A Wood-pecker*	Picus, i, Mártius, *m.*	1 Δρυοκολάπτης, ου, m.
90 *A Wren*	Passer, ĕris, Troglody-tes, *m.*	1 Τρωγλοδύτης, ου, m.

(82) Of this Bird are several sorts, *viz.* The Missel-bird or Shrite, *Turdus viscivorus major*; the Throstle, Song-thrush or Mavis, *Turdus viscivorus minor*; the Red-wing or Swine-pipe, *Turdus Iliacus.*

(84) This is supposed to be the *Meleagris* of the Antients, as also the *Numidica avis.*

(90) This hath been commonly mistaken for the *Regulus* and *Trochilus* of the Antients; but now the *Regulus* is well known to be another Bird less than this, which is also found with us in *England*, but hath no English name that I know.

X.

Of Fishes.	*De Piscibus.*	Περὶ Ἰχθύων.
1 *AN Anchovy*	ENcrasichŏlus, i, *m.*	3 Ἐγκρασίχολος, ου, m.
2 *A Banstickle or Stickle-back*	Pungitius, ii, *m.*	
3 *A Barbel*	Barbus, i, *m.*	
4 *A Blay or Bleak*	Alburnus, i, *m.*	
5 *A Bream*	Cyprīnus, i, latus, *m.*	5 Ἀβραμὶς, ίδος, f.

(2) What the antient Latin name of this Fish was is unknown, *Albertus* named it *Pungitius* from its Prickles, others call it *Spinachia.*

(3,4) The antient Greek names of these Fishes are not known.

(5) The Bream hath been by many taken to be the *Abramis* by reason of the affinity of the words, but *Rondeletius* denies it.

6	*A Button-fish*	Echīnus marinus, *m.*	3 Ἐχῖνος, ου, m.
7	*A Carp*	Cyprīnus, i, *m.*	3 Κυπρῖνος, ου, m.
8	*A Chevin or Chub*	Capĭto, ōnis, *m.*	3 Κέφαλος, άλου, m.
9	*A Cockle*	Concha, æ, striata, *f.*	
10	*A Cod-fish*	Asellus, i, *m.*	3 Ὀνίσκος, ου, m.
11	*A Conger*	Congrus, i, *m.*	3 Κόγγρος, ου, m.
12	*A Crab-fish*	Cancer, cri, *m.*	3 Καρκῖνος, ου, m.
13	*A Cramp-fish*	Torpēdo, ĭnis, *f.*	2 Νάρκη, ης, f.
14	*A Cray-fish*	Cammărus, i, *m.*	3 Κάμμαρος, ου, m.
15	*A Cuttle-fish*	Sepia, æ, *f.*	2 Σηπία, ας, f.
16	*A Dare or Dace*	Leuciscus, i, *m.*	3 Λευκίσκος, ου, m.
17	*A Dog-fish*	Galeus, i, *m.*	3 Γαλεὸς, οῦ, m.
18	*A Dolphin*	Delphīnus, i, *m.*	5 Δελφὶν, ῖνος, m.
19	*A Doree*	Faber, bri, *m.*	3 c. Χαλκεὺς, έως, m.
20	*An Eel*	Anguilla, æ, *f.*	5 Ἔγχελυς, έλυος, f.
21	*An Eel-pout or Burbot*	Mustēla, æ, fluviatilis, *f.*	
22	*A Flair*	Raia, æ, *f.* lævis.	3 Βάτος, ου, m.
23	*A Flounder or Fluke*	Passer, ĕris, *m.*	2 Ψῆττα, ης, f.
24	*A Grayling*	Thymallus, i, *m.*	3 Θύμαλλος, ου, m.
25	*A Gudgeon*	Gobio, ōnis, *m.*	3 Κωβιὸς, οῦ, m.
26	*A Gurnard*	Cucūlus, i, *m.*	5 Κόκκυξ, υγος, m.
27	*An Haddock*	Asĭnus, i, *m.*	3 Ὄνος, ου, m.
28	*An Herring*	Halec, ēcis, *f.* & *n.*	
29	*An Horn-fish or Needle-fish*	Acus, ûs, *f.*	2 Βελόνη, ης, f.
30	*Ling*	Asellus, i, longus, *m.*	3 Ὀνίσκος μακρός.

(9) I see no reason but the Cockle may be called in Latin *Pectunculus*; howbeit I have complied with Naturalists in giving it the general name *Concha* with the distinctive Epithete *striata*.

(28) Although it be generally agreed among latter Naturalists, that our Herring is not the *Halec* of the Antients, yet since it is a general opinion of vulgar Latinists, and neither hath a Herring any antient Latin name, nor *Halec* a known English name, I have let it stand as formerly. That Herring should have no antient Latin or Greek name is no wonder, it being a Fish proper to the Ocean, and so probably unknown to the antient Greeks and Latins.

31	*A Lamprey*	Lampetra, æ, *f.*	
32	*A Limpet*	Patella, æ, *f.*	5 Λεπὰς, άδος, f.
33	*A Lobster*	Astăcus, i, *m.*	3 Ἄστακος, άκου, m.
34	*A Loche*	Cobites, is, fluviatilis, *f.*	
35	*A Long-Oister*	Locusta, æ, marina, *f.*	3 Κάραβος, άβου, m.
36	*A Mackrel*	Scombrus, i, *m.*	3 Σκόμβρος, ου, m.
37	*A Minnew or Pink*	Phoxĭnus, i, *m.*	3 Φόξινος, ίνου, m.
38	*A Mullet*	Mugil, ĭlis, *m.*	3 Κέφαλος, άλου, m.
39	*A Muscle*	Muscŭlus, i, *m.*	5 Μῦς, υός, m.
40	*An Oister*	Ostreum, i, *n.*	3 Ὄστρεον, έου, n.
41	*A Perch*	Perca, æ, *f.*	2 Πέρκη, ης, f.
42	*A Periwincle or Whilk*	Cochlea, æ, *f.*	1 Κοχλίας, ίου, m.
43	*A Pike*	Lucius, ii, *m.*	3 Λύκος, ου, m.
44	*A Plaise*	Passer, ĕris, maculosus *m*	2 Ψῆττα, ης, f.
45	*A Porpus*	Phocæna, æ, *f.*	2 Φώκαινα, ης, f.
46	*A Pourcontrel*	Polypus, i, *m.*	5 Πολύπους, οδος, m.
47	*A Roche*	Rutilus, i, fluviatilis.	
48	*A Ruff*	Cernua, æ, *f.*	
49	*A Salmon*	Salmo, ōnis, *m.*	
50	*A Saw-fish*	Pristis, is, *m.*	1 Πρίστης, ου, m.

(31) This Fish hath been by mistake called *Muræna*; whereas *Muræna* is a much different Fish, common in the Mediterranean Sea, and called at this day *Mourene:* Yet *Rondeletius* will have it to be the *Muræna fluviatilis* of *Athenæus*. However it must by no means be called simply *Muræna* without the distinctive Epithete of *fluviatilis*.

(35) Our Lobster hath been generally but falsly taken for *Locusta marina*; which mistake ought to be rectified.

(38) Heer care must be taken, lest being deceived by the identity of names we take our English *Mullet* to be the *Mullus* of the Antients; much less are we to take our Barbel, a Fish of little account, to be their *Mullus*.

(47) What the antient Latin and Greek names of this Fish were is not certainly known. *Rubellio* and *Erythrinus* are names of a Sea-Fish.

(49) The Salmon being a Fish proper to the Ocean was probably unknown to the antient Greeks, and therefore hath no name in that Language.

51 *A Shad*	Clupea, æ, *f.*	2 Θείσσα, ης, f.
52 *A Sheat-fish*	Silūrus, i, *m.*	3 Σίλουρος, ούρου, m.
53 *A Shark*	Canis, is, Carcharias.	1 Καρχαρίας, ου, m.
54 *A Scate*	Squatĭna, æ, *f.*	2 ʽΡίνη, ης, f.
55 *A Scallop*	Pecten, ĭnis, *m.*	5 Κτεὶς, ενός, m.
56 *A Soal*	Solea, æ, *f.*	2 Βούγλωσσα, ης, f.
57 *A Sturgeon*	Acipenſer, ĕris, *m.*	
58 *A Surmullet*	Mullus, i, *m.*	2 Τρίγλη, ης, f.
59 *A Sword-fish*	Xiphias, æ, *m.*	1 Ξιφίας, ίου, m.
60 *A Smelt*	Violacea, æ, *f.*	
61 *A Shrimp*	Squilla, æ, *f.*	5 Καρὶς, ίδος, f.
62 *A Tench*	Tinca, æ, *f.*	5 Ψύλλων, ωνος, m.
63 *A Thornback*	Raia, æ, clavata.	3 Βάτος, ου, m.
64 *A Turbot*	Rhombus, i, *m.*	3 ʽΡόμβος, ου, m.
65 *A Trout*	Trutta, æ, *f.*	
66 *A Tunny-fish*	Thynnus, i, *m.*	3 Θύννος, ου, m.
67 *A Whale*	Cetus, i, *m.*	1 c. Κῆτος, εος, ους, n.
68 *A Whiting*	Aſellus, i, mollis.	3 ʼΟνίσκος μαλακός.
69 *A Weever*	Draco, onis, *m.*	5 Δράκων, οντος, m.
70 *A Sprat*	Sardīna, æ, *f.*	1 Τριχίας, ίου, m.
71 *A Pilchard*	Harengus minor, *m.*	

(51) The Shad or *Aloſa Rondeletius* will by no means allow to be the *Clupea* of the Latins; it may therefore be called in Latin *Aloſa*, according to the French name.

(54) For the affinity of name I have made *Squatina* the Latin for Scate, though I am not ignorant that *Squatina* is that Fiſh which in ſome places of *England* they call the Monk-fiſh.

(57) Of the antient names of our Sturgeon in Greek and Latin there are almoſt as many Opinions as Authors. That it was the *Acipenſer* of the Latins I think moſt probable: What it was called by the Greeks is not ſo clear. That it was not the *Elops Rondeletius* proves, who would have it to be the *Aſellus Callarias* of *Athenæus.*

(65) This is thought to be the *Salar* of *Auſonius.* What the antient Greek name thereof was is unknown.

XI. Of

XI.

Of Insects.	*De Insectis.*	Περὶ Ἐντόμων.
1 AN *Ant or Emmet*	FOrmica, æ, *f.*	5 Μύρμηξ, ηκος, m.
2 *A Bee*	Apes, is, *f.*	2 Μέλισσα, ης, f.
3 *Honey*	Mel, mellis, *n.*	5 Μέλι, ιτος, n.
4 *An Honey-comb*	Favus, i, *m.*	3 Κηρίον, ίου, n.
5 *An Hive*	Alveāre, is, *n.*	3 Σιμβλὸς, οῦ, m.
6 *A Sting*	Aculeus, i, *m.*	3 Κέντρον, ου, n.
7 *A Drone*	Fucus, i, *m.*	5 Κηφὴν, ῆνος, m.
8 *A Swarm*	Exāmen, ĭnis, *n.*	1 c. Σμῆνος, εος, ους, n.
9 *Bees-wax*	Cera, æ, *f.*	3 Κηρὸς, οῦ, m.
10 *A Humble-Bee*	Bombylius, ii, *m.*	
11 *A Butterfly*	Papilio, ōnis, *m.*	2 Ψυχὴ, ῆς, f.
12 *A Beetle*	Scarabæus, i, *m.*	3 Κάνθαρος, ου, m.
13 *A Caterpillar*	Eruca, æ, *f.*	2 Κάμπη, ης, f.
14 *A Cock-roche*	Blatta, æ, *f.*	2 Σίλφη, ης, f.
15 *A Cricket*	Gryllus, i, *m.*	
16 *A Dragon-fly*	Libella, æ, *f.*	
17 *An Ear-wig*	Forficŭla, æ, *f.*	
18 *A Flea*	Pulex, ĭcis, *m.*	2 Ψύλλα, ης, f.
19 *A Fly*	Musca, æ, *f.*	2 Μυῖα, ας, f.
20 *A Glow-worm*	Cicindēla, æ, *f.*	5 Λαμπουρὶς, ίδος, f.
21 *A Gnat*	Culex, ĭcis, *m.*	5 Κώνωψ, ωπος, m.
22 *A Grass-hopper*	Locusta, æ, *f.*	5 Ἀκρὶς, ίδος, f.
23 *An Hornet*	Crabro, ōnis, *m.*	2 Ἀνθρήνη, ης, f.

(1) This is also called a Pismire.

(16,17) The antient Latin names of these Insects are not known; we have put down the names they are now commonly known and called by by Naturalists. The first is also called *Perla* from the Italian name of it.

(22) *Cicada* by a general mistake in our Schools hath been Englished a Grass-hopper; whereas *Cicada* is an Insect of far different make from the Grass-hopper, proper to hot Countreys, not known in *England*, and having no English name; that usually sits on trees, and sings so loud that it may be heard afar off.

24 *A*

24	*A Horſleech*	Hirūdo, ĭnis, *f.*	2 Βδέλλα, ης, f.
25	*A Louſe*	Pedicŭlus, i, *m.*	5 Φθεὶρ, ρὸς, m.
26	*A Mite*	Syro, ōnis, *m.*	
27	*A Moth*	Tinea, æ, *f.*	5 Σὴς, τὸς, m.
28	*A Maggot*	Eula, æ, *f.*	2 Ἐυλαὶ, ῶν, f.
29	*A Scorpion*	Scorpius, ii, *m.*	3 Σκόρπιος, ίου, m.
30	*A Snail*	Limax, ācis, *com.*	1 Κοχλίας, ου, m.
31	*A Spider*	Aranea, æ, *f.*	2 Ἀράχνη, ης, f.
32	*A Water-Spider*	Tipŭla, æ, *f.*	
33	*A Straw-worm*	Phryganium, ii, *n.*	
34	*A Worm*	Vermis, is, *m.*	5 Σκώληξ, κος, m.
35	*An Earth-worm*	Lumbrīcus, ci, terreſtris, *m.*	5 Ἕλμινς, νθος, f.
36	*A Belly-worm*	Lumbrīcus inteſtinorū.	
37	*A Silk-worm*	Bombyx, ycis, *m.*	5 Βόμβυξ, κος, m.
38	*A Timber-worm*	Terēdo, ĭnis, *f.*	5 Τερηδὼν, όνος, f.
39	*A Wall-louſe or Chinch*	Cimex, ĭcis, *m.*	5 Κόρις, ιδος, f.
40	*A Waſp*	Veſpa, æ, *f.*	5 Σφὴξ, κός, m.
41	*A Weevil*		
42	*A Wood-louſe or Sow*	Aſellus, i, *m.*	3 Ὀνίσκος, ου, m.
43	*A Breez or Gad-fly*	Aſilus, i, *m.*	3 Οἶστρος, ου, m.
44	*A Tick*	Ricĭnus, i, *m.*	5 Κρότων, ωνος, m.
45	*A Nit*	Lens, dis, *f.*	5 Κόνις, ιδος, f.

(36) Of Belly-worms there be three uſual ſorts, 1. The round ones called *Teretes*. 2. The flat ones called *Lati*. 3. Thoſe called *Aſcarides*; for *Aſcarides* is not the general name of all Belly-worms.

XII.

Of the Parts of Mans Body.	*De Partibus Humani Corporis.*	Περὶ τῶν Ἀνθρωπίνου Σώματος Μερῶν.

In ſetting down the Parts of Mans Body, I ſhall follow the Diviſion of the late L. Bp of *Cheſter* in his Book of the Univerſal Character; which is into Homogeneous or ſimilar, and Heterogeneous or diſſimilar Parts; the Homogeneous being ſubdivided into conteining and conteined, the Heterogeneous into external and internal.

I. Of

I. *Of Homogeneous Parts conteining.*

The Body	Corpus, ŏris, *n.*	5 Σῶμα, τος, n.
A Member	Membrum, i, *n.*	1c. Μέλος, εος, ους, n
A Limb	Artus, ûs, *m.*	3 Κῶλον, ου, n.
A Bone	Os, ossis, *n.*	3 Ὀστέον, οῦν, έου, οῦ, n.
A Gristle	Cartilāgo, ĭnis, *f.*	5 Ἀδὴν, ένος, m.
A Sinew	Nervus, i, *m.*	3 Νεῦρον, ου, n.
A Vein	Vena, æ, *f.*	5 Φλὲψ, βός, f.
An Artery	Arteria, æ, *f.*	2 Ἀρτηρία, ας, f.
Flesh	Caro, nis, *f.*	5 Σάρξ, κός, f.
A Muscle	Muscŭlus, i, *m.*	5 Μῦς, υός, m.
The Skin	Cutis, is, *f.*	5 Χρὼς, ωτός, m.
A Membrane or Film	Membrāna, æ, *f.*	5 Ὑμὴν, ένος, m.
A Fibre	Fibra, æ, *f.*	5 Ἶς, ἰνός, f.
The Hair of the Head	Capillus, i, *m.*	5 Θρὶξ, τριχός, f.
A curled Lock	Cincinnus, i, *m.*	5 Βόστρυξ, υχος, m.

II. *Of Homogeneous Parts conteined.*

The Brain	Cerĕbrum, i, *n.*	3 Ἐγκέφαλος, άλου, m.
Marrow	Medulla, æ, *f.*	3 Μυελὸς, οῦ, m.
Fat or Grease	Adeps, ĭpis, *dub.*	2 Πιμελὴ, ῆς, f.
Tallow or Suet	Sebum, i, *n.*	5 Στέαρ, ατος, n.
Blood	Sanguis, ĭnis, *m.*	5 Αἷμα, ατος, n.
Milk	Lac, lactis, *n.*	5 Γάλα, ακτος, n.
Gall	Fel, fellis, *n.* }	2 Χολὴ, ῆς, f.
Choler	Bilis flava, *f.* }	
Melancholy	atra, *f.*	2 Μελαγχολία, ας, f.
Flegm	Pituīta, æ, *f.*	5 Φλέγμα, ατος, n.
Snot	Mucus, i, *m.*	2 Μύξα, ης, f.
Urin	Urīna, æ, *f.*	3 Οὖρον, ου, n.
Dung	Stercus, ŏris, *n.*	3 Κόπρος, ου, m.
Sweat	Sudor, ōris, *m.*	5 Ἱδρὼς, ῶτος, m.
Spittle	Salīva, æ, *f.*	3 Πτύελον, ου, n.

III. *Of*

III. *Of external Heterogeneous Parts.*

The Head	Caput, ĭtis, *n.*	2 Κεφαλὴ, ῆς, f.
The Fore-part } *of the Head*	Sincĭput, ĭtis, *n.*	3 Βρεγμὸς, οῦ, m.
The Hinder-part } *of the Head*	Occĭput, ĭtis, *n.*	3 Ἰνίον, ίου, n.
The Crown	Vertex, ĭcis, *m.*	2 Κορυφὴ, ῆς, f.
The Face	Facies, ei, *f.*	2 c. Ὄψις, εως, f.
The Countenance	Vultus, ûs, *m.*	3 Πρόσωπον, ώπου, n.
The Skull	Cranium, ii, *n.*	3 Κρανίον, ίου, n.
The Forehead	Frons, tis, *f.*	3 Μέτωπον, ώπου, n.
The Eye-brow	Supercilium, ii, *n.*	3 Ἐπισκύνιον, ίου, n.
The Eye-lid	Palpĕbra, æ, *f.*	3 Βλέφαρον, άρου, n.
The Eye	Ocŭlus, i, *m.*	3 Ὀφθαλμὸς, οῦ, m.
The Sight of the Eye	Pupilla, æ, *f.*	2 Κόρη, ης, f.
The White of the Eye	Albūgo, ĭnis, *f.*	5 Λεύκωμα, ατος, n.
The Corner of the Eye	Hirquus, i, *m.*	3 Κανθὸς, οῦ, m.
The Hairs of the Eye-lids	Cilia, orum, *n.*	5 Βλεφαρὶς, ίδος, f.
The Ear	Auris, is, *f.*	5 Οὖς, ὠτός, n.
The Temples of the Head	Tempŏra, um, *n.*	3 Κρόταφος, άφου, m.
The Nose	Nasus, i, *m.*	5 Μυκτὴρ, ῆρος, m.
The Nosthrils	Nares, ium, *f.*	5 Ῥὶς, ῥινός, f.
A Cheek	Gena, æ, *f.*	2 Παρειὰ, ᾶς, f.
The Lip	Labium, ii, *n.*	1 c. Χεῖλος, εος, ους, n.
The Mouth	Os, oris, *n.*	5 Στόμα, ατος, n.
The Palat or Roof of the mouth	Palātum, i, *n.*	2 Ὑπερῴα, ας, f.
The Gums	Gingīva, æ, *f.*	3 Οὖλα, pl. οὔλων, n.
A Tooth	Dens, tis, *m.*	5 Ὀδοὺς, όντος, m.
The Jawbone	Maxilla, æ, *f.*	5 Σιαγὼν, όνος, f.
The Fore-teeth	Incisōres, um, *m.*	1 Τομῆται, τῶν, m.
The Jaw-teeth	Molāres, ium, *m.*	1 Μυλῖται, τῶν, m.
The Tongue	Lingua, æ, *f.*	2 Γλῶσσα, ης, f.
The Chin	Mentum, i, *n.*	3 Γένειον, είου, n.
The Beard	Barba, æ, *f.*	5 Πώγων, ωνος, m.
The Neck	Collum, i, *n.*	3 Τράχηλος, ήλου, m.
The Throat	Guttur, is, *n.*	3 Βρόγχος, ου, m.

The

The Shoulder	Humĕrus, i, *m.*	3 Ὦμος, ὤμȣ, m.
The Shoulder-blade	Scapŭla, æ, *f.*	2 Ὠμοπλάτη, ης, f.
The Back	Dorſum, i, *n.*	3 Νῶτος, ȣ, m.
The Breaſt	Pectus, ŏris, *n.*	3 Στέρνον, ȣ, n.
A Pap	Mamma, æ, *f.*	3 Μαστὸς, ȣ̃, m.
A Nipple	Papilla, æ, *f.*	2 Θήλη, ης, f.
The Boſom	Sinus, ûs, *m.*	3 Κόλπος, ȣ, m.
An Udder	Uber, ĕris, *n.*	5 Οὖθαρ, ατος, n.
A Side	Latus, ĕris, *n.*	3 Πλευρὸν, ȣ̃, n.
The Backbone	Spina, æ, *f.*	2 c. Ῥάχις, εως, f.
A Rib	Coſta, æ, *f.*	2 Πλευρὰ, ᾶς, f.
A Loin	Lumbus, i, *m.*	5 Ὀσφὺς, ύος, f.
The Belly	Venter, tris, *m.*	5 Γαστὴρ, έρος, τρός, f.
The Navel	Umbilīcus, ci, *m.*	3 Ὀμφαλὸς, λȣ̃, m.
The Hucklebone	Coxa, æ, *f.*	2 Κοτύλη ης, f.
The Hip	Coxendix, ĭcis, *f.*	3 Ἰσχίον, ίȣ, n.
The Flank	Ilia, ium, *n.*	5 Κενεὼν, ῶνος, m.
The Thigh	Femur, ŏris, *n.*	3 Μηρὸς, ȣ̃, n.
The Knee	Genu, *indec. n.*	5 Γόνυ υος, n.
The Ham	Poples, ĭtis, *m.*	2 Ἰγνύα, ας, f.
The Leg	Crus, crūris, *n.*	1 c. Σκέλος, εος ȣς, n.
The Shin	Tibia, æ, *f.*	3 Ἀντικνήμιον, ίȣ, n.
The Calf of the Leg	Sura, æ, *f.*	2 Γαστροκνημία ας, f.
The Ancle	Malleŏlus, i, *m.*	3 Σφύριον, ίȣ, n.
The Foot	Pes, pedis, *m.*	5 Πȣ̃ς, ποδὸς, m.
A Toe	Digĭtus pedis, *m.*	3 Δάκτυλος ποδὸς, m.
The great Toe	Hallux, ŭcis, *m.*	5 Ἀντίχειρ, ερος, m.
The Heel	Calx pedis *com.*	2 Πτέρνα, ης, f.
The Sole of the Foot	Planta, æ, pedis, *f.*	5 Πέλμα ατος, n.
The Arm	Brachium, ii, *n.*	5 Βραχίων ονος, m.
The Arm-pit	Axilla, æ, *f.*	2 Μασχάλη, ης, f.
The Elbow	Cubĭtus, i, *m.*	5 Πῆχυς, εως, m.
The Wriſt	Carpus, i, *m.*	3 Καρπὸς, ȣ̃, m.
The Hand	Manus, ûs, *f.*	5 Χεὶρ, χειρὸς, f.
The Right Hand	Dextra manus, *f.*	2 Δεξιὰ ᾶς, f.

The Left Hand	Sinistra, æ, *f.*	2 Ἀριστερὰ, ᾶς, f.
The Palm of the Hand	Palma, æ, *f.*	2 Παλάμη, ης, f.
The Back of the Hand	Metacarpium, ii, *n.*	5 Ὀπισθέναρ, αρος, n.
The Hollow of the Hand	Vola, æ, *f.*	5 Θέναρ, αρος, n.
The Fist	Pugnus, i, *m.*	5 Δρὰξ, ακός, m.
A Finger	Digĭtus, i, *m.*	3 Δάκτυλος, ύλου, m.
The Fore-finger	Index, ĭcis, *m.*	3 Λιχανὸς, οῦ, m.
The Middle-finger	Verpus, i, *m.*	3 Ψωλὸς, οῦ, m.
The Ring-finger	Annulāris, is, *m.*	1 Δακτυλιώτης, ου, m.
The Little-finger	Auriculāris, is, *m.*	1 Ὠτίτης δάκτυλος, m.
The Thumb	Pollex, ĭcis, *m.*	5 Ἀντίχειρ, ερός, m.
A Knuckle	Condylus, i, *m.*	3 Κόνδυλος, ύλου, m.
A Nail	Unguis, is, *m.*	5 Ὄνυξ, χος, m.
A Joint	Artĭcŭlus, i, *m.*	3 Ἄρθρον, ου, n.

IV. *Of internal Heterogeneous Parts.*

The Gullet	Gula, æ, *f.*	3 Οἰσόφαγος, άγου, m.
The Wind-pipe	Aspera Arteria, æ, *f.*	2 Τραχεῖα, ας ἀρτηρία f.
The Lungs or Lights	Pulmo, ōnis, *m.*	5 Πνεύμων, ονος, m.
The Breath	Spirĭtus, ûs, *m.*	5 Πνεῦμα, ατος, n.
The Midriff	Diaphragma, ătis, *n.*	5 Διάφραγμα, ατος, n.
The Heart	Cor, dis, *n.*	2 Καρδία, ας, f.
The Stomach	Ventricŭlus, i, *m.*	3 Γαστρίδιον, ίου, n.
The mouth of the Stomach	Stomăchus, i, *m.*	3 Στόμαχος, άχου, m.
The Bowels	Viscĕra, um, *n.*	3 Σπλάγχνα, ων, n.
The Small-guts	Lactes, ium, *f.*	5 Χολάδες, ων, f.
A Gut	Intestīnum, i, *n.*	3 Ἔντερον, έρου, n.
The Liver	Jecur, ŏris, *n.*	5 Ἧπαρ, ατος, n.
The Spleen or Milt	Lien, ēnis, *m.*	5 Σπλὴν, σπληνός, m.
The Mesentery	Mesenterium, ii, *n.*	3 Μεσεντέριον, ίου, n.
The Caul	Omentum, i, *n.*	3 Ἐπίπλοον, όου, n.
A Kidney	Ren, rēnis, *m.*	3 Νεφρὸς, οῦ, m.
The Bladder	Vesīca, æ, *f.*	1 c. Κύστις, εως, f.

XIII.

Of some Accidents of the Body.	*De quibusdam Corporis Accidentibus.*	Περὶ τῶν τοῦ Σώματος Συμβεβηκότων.
THe Look	ASpectus, ûs, *m.*	2 c. ΠΡόσοψις εως, f.
Paleness	Pallor, ōris, *m.*	5 Ὠχρότης ητος, f
Beauty	Pulchritūdo, ĭnis, *f.*	1 c. Κάλλος, εος ους, n.
Fair or handsom	Formōsus, i, *m.*	3 Καλὸς, οῦ, m.
Ill-favoured or ugly	Deformis, is, *m.*	3 Αἰσχρὸς, οῦ, m.
Gross	Crassus, i, *m.* obesus.	5 Παχὺς, έος, m.
Slender	Gracĭlis, is, *m.*	3 Ἰσχνὸς, οῦ, m.
Fat	Pinguis, is, *m.*	1 c. Πιμελὴς, έος οῦς, m.
Lean	Macer, cri, *m.*	3 Λεπτὸς οῦ, m.
Sound or Hail	Sanus, i, *m.*	1 c. Ὑγιὴς έος οῦς, com.
Lusty	Valĭdus, i, *m.*	3 Ἰσχυρὸς οῦ, m. (κης.
Tall	Procērus, i, *m.*	3 Μακρὸς οῦ, m. Εὐμή-
Low	Humĭlis, is, *m.*	3 Μικρὸς τῇ ἡλικίᾳ.
A Giant	Gigas, antis, *m.*	5 Γίγας, αντος, m.
A Dwarf	Nanus, i, *m.*	3 Νάνος, ου, m.
Jolt-headed	Capĭto, ōnis, *m.*	3 Μεγαλοκέφαλος, άλου, m.
Bald	Calvus, i, *m.*	3 Φαλακρὸς οῦ, m.
Curl'd	Crispus, i, *m.*	5 Οὐλόθριξ, τριχος, m.
Blind	Cæcus, i, *m.*	3 Τυφλὸς, οῦ, m.
Purblind	Myops, ōpis, *m.*	5 Μύωψ, ωπος, com.
One-eyed	Luscus, i, *m.*	3 Ἑτερόφθαλμος, ου, m.
Blear-eyed	Lippus, i, *m.*	5 Λημῶν, ῶντος, m.
Squint-eyed	Strabo, ōnis, *m.*	3 Στραβὸς, οῦ, m.
Deaf	Surdus, i, *m.*	3 Κωφὸς, οῦ, m.
Stammering	Balbus, i, *m.*	3 Ψελλὸς, οῦ, m.
Lisping	Blæsus, i, *m.*	3 Τραυλὸς, οῦ, m.
Toothless	Edentŭlus, i, *m.*	5 Ἀνόδους, οντος, m.
Dumb	Mutus, i, *m.*	3 Ἄφωνος, ώνου, m.

Long-

Long-tongued	Linguax, ācis, *m.*	1 c. Γλωσσώδης, εος, ους, c.
Great-nosed	Naſo, ōnis, *m.*	5 Μεγαλόρριν, ινος, m.
Blubber-lipped	Labeo, ōnis, *m.*	5 Χείλων, ωνος, m.
Crump-ſhouldered	Gibbōſus, i, *m.*	3 Κυρτὸς, οῦ, m.
Left-handed	Scævus, i, *m.*	3 Σκαιὸς, οῦ, m.
Gor-bellied	Ventricōſus, i, *m.*	5 Προγάστωρ, ορος, m.
Bow-legged	Valgus, i, *m.*	3 Βλαισὸς, οῦ, m.
Splay-footed	Plancus, i, *m.*	5 Πλατύπους, οδος, m.
Lame	Claudus, i, *m.*	3 Χωλὸς, οῦ, m.
Maimed	Mancus, i, *m.*	3 Πηρὸς, οῦ, m.
Sleep	Somnus, i, *m.*	3 Ὕπνος, ου, m.
Watching	Vigilia, æ, *f.*	2 Ἀγρυπνία, ας, f.
A Dream	Inſomnium, ii, *n.*	3 Ἐνύπνιον, ίου, n.
Panting	Anhelītus, ûs, *m.*	5 Ἆσθμα, ατος, n.
The Hiccough	Singultus, ûs, *m.*	5 Λύγξ, γγός, f.
Sneezing	Sternutatio, ōnis, *f.*	3 Πταρμὸς, οῦ, m.
Yawning	Oſcitatio, ōnis, *f.*	2 Χάσμη, ης, f.
Stretching	Pandiculatio, ōnis, *f.*	5 Σκορδίνημα, ατος, n.
Hunger	Fames, is, *f.*	3 Λιμὸς οῦ, m.
Thirſt	Sitis, is, *f.*	2 Δίψα, ης, f.
Loathing	Faſtidium, ii, *n.*	2 Ναυτία, ας, f.
The Voice	Vox, vōcis, *f.*	2 Φωνὴ, ῆς, f.
Speech	Sermo, ōnis, *m.*	3 Λόγος, ου, m.
Laughter	Riſus, ûs, *m.*	5 Γέλως, ωτος, m.
Weeping	Fletus, ûs, *m.*	3 Κλαυθμὸς, οῦ, m.
Whiſpering	Suſurrus, i, *m.*	3 Ψιθυρισμὸς, οῦ, m.
Singing	Cantus, ûs, *m.*	2 Ὠδὴ, ῆς, f.
A Sigh	Suſpirium, ii, *n.*	3 Στεναγμὸς, οῦ, m.
Snorting	Rhoncus, i, *m.*	3 Ῥόγχος, ου, m.
Bluſhing	Rubor, ōris, *m.*	1 c. Ἔρευθος, εος, ους, n.
Quaking or Trembling	Tremor, ōris, *m.*	3 Τρόμος, ου, m.
Gray-hairs	Canities, ei, *f.*	5 Πολιότης, ητος, m.
A Wrinkle	Ruga, æ, *f.*	5 Ῥυτὶς, ίδος, f.

XIV. Of

XIV.

Of Diseases.	*De Morbis.*	Περὶ Νοσημάτων.
A Disease	Morbus, i, *m.*	3 Νόσος, ου, f.
Sick	Ægrōtus, i, *m.*	3 Ἄῤῥωστος, ου, com.
Sickness	Valetūdo adversa.	2 Ἀῤῥωστία, ας, f.
Weak	Infirmus, i, *m.*	1 c. Ἀσθενὴς, έος, οῦς, c.
Pain	Dolor, ōris, *m.*	1 c. Ἄλγος, εος, ους, n.
A Wound	Vulnus, ĕris, *n.*	5 Τραῦμα, ατος, n.
A Bruise	Contusio, ōnis, *f.*	5 Θλάσμα, ατος, n.
A Stroke	Plaga, æ, *f.*	2 Πληγὴ, ῆς, f.
A Sore or Ulcer	Ulcus, ĕris, *n.*	1 c. Ἕλκος, εος, ους, n.
A Swelling	Tumor, ōris, *m.*	3 Ὄγκος, ου, m.
A Gangrene	Gangræna, æ, *f.*	2 Γάγγραινα, ης, f.
A Felon	Furuncŭlus, i, *m.*	3 Δοθίον, ου, n.
A Tetter or Ringworm	Herpes, ētis, *m.*	5 Ἕρπης, ητος, m.
A Blister	Vesicŭla, æ, *f.*	2 Φύσκα, ης, & Φύσιγξ.
A Wheal	Papŭla, æ, *f.*	2 Φλύκταινα, αίνης, f.
A Whitblow	Paronychia, æ, *f.*	2 Παρωνυχία, ας, f.
A Pimple	Pustŭla, æ, *f.*	2 Φλύκταινα, ης, f.
Freckles	Lentīgo, ĭnis, *f.*	3 Φακὸς, οῦ, m.
A Scald-Head	Tinea capitis.	
Dandroof	Furfures, um, *m.*	2 Πιτυρίασις, άσεως, f.
St. Anthony's Fire	Erysipĕlas, ătis, *n.*	5 Ἐρυσίπελας, ατος, n.
The Leprosie	Lepra, æ, *f.*	2 Λέπρα, ας, f.
The Itch	Scabies, ei, *f.*	2 Ψώρα, ας, f.
Itching	Prurītus, ûs, *m.*	3 Κνησμὸς, οῦ, m.
A Scab of a wound or sore	Crusta, æ, *f.*	2 Ἐσχάρα, ας, f.
A Scar	Cicātrix, īcis, *f.*	2 Οὐλὴ, ῆς, f.
Hard-Skin or Brawniness	Callus, i, *m.*	3 Τύλος, ου, m.
The Plague	Pestis, is, *f.*	3 Λοιμὸς, οῦ, m.
A Fever	Febris, is, continua, *f.*	3 Πυρετὸς, οῦ, m. Συνεχὴς

An

An Ague	Febris intermittens.	3 Πυρετὸς διαλείπων.
The third Ague or Quartan	Febris quartana.	3 Τεταρταῖος, ου, m.
The Tertian or each day Ague	Febris tertiana.	3 Τριταῖος, ου, m.
The Quotidian or every day Ague	Febris quotidiana.	3 Ἀμφημερινὸς, οῦ, m.
The Small Pox	Variolæ, ārum, *f.*	
The Measles	Morbilli, ōrum, *m.*	
The French-Pox	Lues, is, Venerea.	
The Consumption	Tabes, is, *f.*	2 c. Φθίσις, εως, f. Μαρασμὸς, οῦ, m.
The Dropsie	Hydrops, ōpis, *m.*	5 Ὕδρωψ, πος, m.
The Falling-Sickness	Epilepsia, æ, *f.*	2 Ἐπιληψία, ας, f.
The Apoplexy	Apoplexia, æ, *f.*	2 Ἀποπληξία, ας, f.
The Lethargy	Lethargus, i, *m.*	3 Λήθαργος, ου, m.
Drowsiness	Veternus, i, *m.*	5 Νωθρότης, ητος, f.
Amazedness	Stupor, ōris, *m.*	2 c. Ἔκπληξις, εως, f.
The Palsie	Paralysis, is, *f.*	2 c. Παράλυσις, εως, f.
Numbness	Torpor, ōris, *m.*	2 c. Νάρκωσις, εως, f.
The Cramp	Spasmus, i, *m.*	3 Σπασμὸς, οῦ, m.
The Night-mare	Incŭbus, i, *m.*	1 Ἐφιάλτης, ου, m.
Bedrid	Clinĭcus, i, *m.*	3 Κλινικὸς, οῦ, m.
Dizziness	Vertīgo, ĭnis, *f.*	3 Δῖνος, ου, m.
Frensie	Phrenĕsis, is, *f.*	5 Φρενῖτις, ιδος, f.
Madness	Insania, æ, *f.*	2 Μανία, ας, f.
Dotage	Delirium, ii, *n.*	2 Παραφροσύνη, ης, f.
The Head-ach	Cephalalgia, æ, *f.*	2 Κεφαλαλγία, ας, f.
The Megrim	Hemicrania, æ, *f.*	2 Ἡμικρανία, ας, f.
The Tooth-ach	Odontalgia, æ, *f.*	2 Ὀδονταλγία, ας, f.
A Cold	Gravēdo, ĭnis, *f.*	3 Κόρυζα, ης, f.
A Cough	Tussis, is, *f.*	5 Βὴξ, χός, f.
A Catarrh	Catarrhus, i, *m.*	3 Κατάῤῥος, ου, m.
The Chin-cough	Catarrhus ferinus sive Tussis convulsiva.	
Hoarsness	Raucēdo, ĭnis, *f.*	1 c. Βράγχος, εος, ους, n.
The Rickets	Rachitis, ĭdis, *f.*	
The Squinancy	Angīna, æ, *f.*	2 Κυνάγχη, ης, f.
Rheum	Rheuma, ătis, *n.*	5 Ῥεῦμα, ατος, n.

The

The Tissick	Asthma, ătis, *n.*	5 Ἆσθμα, ατος, n.
Fainting	Languor, ōris, *m.*	2 c. Πάρεσις, εως, f.
Swouning	Lipothymia, æ, *f.*	2 Λειποθυμία, ας, f.
The Pleurisie	Pleurītis, itĭdis, *f.*	5 Πλευρῖτις, ιδος, f.
Heart-burning	Cardialgia, æ, *f.*	2 Καρδιαλγία, ας, f.
Womens Longing	Pica, æ, *f.*	2 Κίττα, ης, f.
The Green-Sicknes	Chlorōsis, is, *f.*	2 c. Χλώρωσις, εως, f.
The Jaundise	Ictĕrus, i, *m.*	3 Ἴκτερος, έρου, m.
A Tympany	Tympanītes Hydrops.	1 Τυμπανίτης, ου, m.
The Scurvy	Scorbūtus, i, *m.*	2 Στομακάκη, ης, f.
The Colick	Colĭcus dolor, *m.*	2 Κωλικὴ διάθεσις, f.
A Loosenes	Diarrhœa, æ, *f.*	2 Διάῤῥοια, ας, f.
Costivenes	Alvus astricta seu dura	2 Κοιλία ξηρά, f.
The Bloody Flix	Dysenteria, æ, *f.*	2 Δυσεντερία, ας, f.
The Stone	Calcŭlus, i, *m.*	2 c. Λιθίασις, άσεως, f.
Chafing or Fretting	Intertrīgo, ĭnis, *f.*	5 Παράτριμμα, ατος, n.
The Strangury	Stranguria, æ, *f.*	2 Στραγγουρία, ας, f.
A Rupture or Burstnes	Hernia, æ, *f.*	2 Κήλη, ης, f.
The Mother	Hysterica, æ, Passio, *f.*	5 Πνὶξ, γὸς, ὑστερική, f.
The Gout	Arthrītis, itĭdis, *f.*	5 Ἀρθρῖτις, ιδος, f.
The Gout of the Foot	Podăgra, æ, *f.* *which word is also generally used for the Gout in any part.*	
The Running Gout	Rheumatismus, i, *m.*	3 Ῥευματισμὸς, οῦ, m.
The Piles	Hæmorrhoides, um, *f.*	5 Αἱμοῤῥοΐδες, ων, f.
The Kings-Evil	Strumæ, arum, *f.*	5 Χοιράδες, ων, f.
A Wolf or Cancer	Cancer, cri, *m.*	5 Καρκίνωμα, τος, n.
A Wart	Verrūca, æ, *f.*	5 Ἀκροχορδὼν, όνος, m.
A Mole	Nævus, i, *m.*	3 Σπῖλος, ου, m.
A Corn	Clavus, i, pedis, *m.*	3 Ἧλος, ου, m.
A Kibe or Chilblain	Pernio, ōnis, *f.*	3 Χίμεθλον, ου, n.
Hang-nails	Reduvia, æ, *f.*	2 Παρωνυχία, ας, f.
A Fit	Paroxysmus, i, *m.*	3 Παρόξυσμος, ου, m.
A Physician	Medĭcus, i, *m.*	3 Ἰατρὸς οῦ, m.
Physick	Medicīna, æ, *f.*	2 Ἰατρικὴ, ῆς, f.
A Medicin	Medicamentum, i, *n.*	3 Φάρμακον, άκου, n.

A Re-

A Remedy	Remedium, ii, *n.*	2 c. Ἄκος, εος, ους, n.
An Apothecary	Pharmacopôla, æ, *m.*	1 Φαρμακοπώλης, ου, m.
A Chirurgion	Chirurgus, i, *m.*	3 Χειρουργὸς, οῦ, m.
Bloud-letting	Phlebotomia, æ, *f.*	2 Φλεβοτομία, ας, f.
A Vomit	Vomĭtus, ûs, *m.*	3 Ἔμετος, ου, m.
Purging	Purgatio, ōnis, *f.*	2 c. Κάθαρσις, εως, f.
A Clyster	Enĕma, ătis, *n.*	5 Κλυστὴρ, ῆρος, m.
A Potion	Potio, ōnis, *f.*	5 Πόμα, ατος, n.
A Syrup	Syrūpus, i, *m.*	3 Σιράπιον, ίου, n.
A Pill	Pilŭla, æ, *f.*	3 Καταπότιον, ίου, n.
A Plaster or Salve	Emplastrum, i, *n.*	3 Ἔμπλαστρον, ου, n.
A Sear-cloth	Cerātum, i, *n.*	3 Κηρωτὸν, οῦ, n.
A Tent	Turunda, æ, *f.* Penicillus	3 Μοτὸς, οῦ, m.
Ointment	Unguentum, i, *n.*	3 Μύρον, ου, n.
A Wen	Struma, æ, *f.*	5 Χοιρὰς, άδος, f.

XV.

Of Meat.	*De Cibo.*	Περὶ Βρωμάτων.
A Caterer	Obsonātor, ōris, *m.*	1 Ὀψώνης, ου, m.
Food	Victus, ûs, *m.*	2 Τροφὴ, ῆς, f.
A Meal	Refectio, ōnis, *f.*	
A Break-fast	Jentacŭlum, i, *n.*	5 Ἀκράτισμα, ατος, n.
A Dinner	Prandium, ii, *n.*	3 Ἄριστον, ου, n.
A Beaver	Merenda, æ, *f.*	3 Προδείπνον, ου, n.
A Supper	Cœna, æ, *f.*	3 Δεῖπνον, ου, n.
A Feast	Convivium, ii, *n.*	3 Συμπόσιον, ίου, n.
A Guest	Convīva, æ, *m.*	3 Σύνδειπνος, οίνου, m.
A Banquet	Epŭlum, i, *n.*	2 Εὐωχία, ας, f.
Junkets or Sweet-meats	Bellaria, orum, *n.*	5 Τραγήματα, άτων, n.
The first Course	Prima mensa, æ, *f.*	2 Τράπεζα πρώτη.
The second Course	Mensæ secundæ.	2 Τράπεζαι δεύτεραι. Ἐπιδορπὶς, ίδος, f.

A Mess

A Mess	Fercŭlum, i, *n.*	2 c. Παράθεσις, εως, f.
Grace	Gratiarum actio.	2 Εὐχαριστία, ας, f.
Dainties	Dapes, um, *f.*	2 Τράπεζα πολυτελής.
A Table	Mensa, æ, *f.*	2 Τράπεζα, έζης, f.
A Table-Cloth	Mappa, æ, *f.*	2 Ὀθόνη, ης, f.
A Napkin or Towel	Mantelium, ii, *n.*	3 Χειρόμακτρον, ου, n.
An Ewer	Aquālis, is, *m.*	3 Πρόχοος, όου, f.
A Bason	Pelvis, is, *m.*	5 Νιπτὴρ, ῆρος, m.
A square Trencher	Quadra, æ, *f.*	5 Πίναξ, ακος, m.
A round Trencher or Plate	Orbis, is, *m.*	3 Κύκλος, ου, m.
A Salt-seller	Salīnum, i, *n.*	3 Ἁλοδοχεῖον, ου, n.
Salt	Sal, salis, *m.*	5 Ἅλς, λός, m.
Bread / *A Loaf*	Panis, is, *m.*	3 Ἄρτος, ου, m.
A Morsel	Buccea, æ, *f.*	3 Ψωμὸς, οῦ, m.
Fine Flower	Pollen, ĭnis, *m.*	2 Παιπάλη, ης, f.
Unlevened Bread	Azymus, i, *m.* (Panis)	3 Ἄζυμος, ύμου, m. (Ἄρτος)
Levened Bread	Fermentātus, i, *m.* (Panis)	1 Ζυμίτης, ου, m. (Ἄρτος)
Manchet or fine white Br.	Siligineus, i, *m.* (Panis)	1 Σιλιγνίτης, ου, m. (Ἄρτος)
Temse Bread	Similaceus, i, *m.* (Panis)	1 Σεμιδαλίτης, ου, m (Ἄρτος)
Houshold Bread	Cibarius, i, *m.* (Panis)	3 Αὐτόπυρος, ου, m (Ἄρτος)
Bisket	Biscoctus, i, *m.* (Panis)	3 Ναυτικὸς, οῦ, m. (Ἄρτος)
The Crust	Crusta, æ, *f.*	5 Πλὰξ, κός, f.
The Crum	Medulla, æ, *f.* panis	3 Μύελος, έλου, m.
A Crum of Bread	Mica, æ, *f.*	3 Ψιχίον, ου, n.
A Wafer	Crustŭlum, i, *n.* Libū, i, *n.*	3 Πλακούντιον ίου, n.
A Simnel	Torta, æ, *f.*	5 Κολλυρὶς, ίδος, f.
A Pancake or Fritter	Lagănum, i, *n.*	3 Λάγανον, ου, n.
A Cake	Placenta, æ, *f.*	5 Πλακοῦς οῦς, όεντος οῦντος
A Bread-basket	Canistrum, i, *n.*	5 Κάνης, ητος, m.
A Knife	Culter, tri, *m.*	2 Μαχαιρὶς, ίδος, f.
An Edge	Acies, ei, *f.*	5 Στόμα, ατος, n. Ἀκμὴ f
A Carver	Structor, ōris, *m.*	3 Σιτοτόμος, ου, m.
A Dish	Discus, i, *m.*	3 Δίσκος, ου, m.
A Porringer	Catillus, i, *m.*	3 Τρυβλίδιον, ίου, n.

A Platter

A Platter	Patĭna, æ, f.	2 Λεκάνη, ης, f. Παροψὶς, (ιδος, f.
A Sawcer	Acetabŭlum, i, n.	3 Ὀξύβαφον, ου, n.
A Spoon	Cochleāre, is, n.	2 Τορύνη, ης, f. Ζωμήρυσις (εως, f.
Pottage or Broth	Juſcŭlum, i, n.	3 Ζωμὸς, οῦ, m.
Water-gruel	Pulmentum, i, n.	
Cream	Flos lactis, m.	5 Ἀφρόγαλα, ακτος, n.
Milk-meats	Lacticinia, orum, n.	3 Γαλάκτινα, ων, n.
A Posset	Zythogăla, actos, n.	5 Ζυθόγαλα, ακτος, n.
A Syllibub	Oxygăla, actos, n.	5 Ὀξύγαλα, ακτος, n.
Butter	Butyrum, i, n.	3 Βούτυρον, ου, n.
Cheese	Caſeus, i, m.	3 Τυρὸς, οῦ, m.
A Custard	Artogăla, ctos, n.	5 Ἀρτόγαλα, ακτος, n.
Whey	Serum, i, lactis, n.	3 Ὀῤῥὸς, οῦ, m.
Flesh	Caro, nis, f.	5 Σὰρξ, κός, f.
Beef	Caro Bovina, æ, f.	Κρέας Βόειον, είου.
Veal	Caro Vitulina	Κρέας Μόσχειον, είου.
Mutton	Caro Ovina	Κρέας Προβάτειον, είου.
Lamb	Caro Agnina	Κρέας Ἄρνειον, είου.
Pork	Caro Porcina	Κρέας Χοίρειον, είου.
Venison	Caro Ferina	Κρέας Θήρειον, είου.
Bacon	Lardum, i, n.	2 Πιμελὴ χοιρεία.
A Flitch	Succidĭa, æ, f.	2 Πέρνα, ης, f.
A Gammon of Bacon	Petăſo, ōnis, m.	5 Πετασὼν, ῶνος, m.
Brawn	Callum aprugnum	
Rost Meat	Caro Aſſa, æ	Κρέας Ὀπτὸν, οῦ.
Boil'd Meat	Caro Elixa, æ	Κρέας Ἑφθὸν, οῦ.
Bak'd Meat	Caro Coctilis, is	
Broil'd Meat	Caro Toſta, æ	Κρέας Ὠπτημένον, ου.
Fried Meat	Caro Frixa, æ	Κρέας Φρυκτὸν, οῦ.
Stewed Meat	Caro Jurulenta, æ	Κρέας Ζωμῶδες.
Carbonadoed Meat	Carbonella, æ, f.	5 Ἀπανθράκισμα, τος, n.
Minc'd Meat	Minūtal, is, n.	5 Περίκομμα, τος, n.
Meat cut in Gobbets	Tomacŭlum, i, n.	3 Τεμάχιον, ίου, n.
A Gut-Pudding	Farcīmen, ĭnis, n.	5 Ἀλλᾶς, ᾶντος, m.
A Pudding	Fartum, i, n.	

A Pud.

A Pudding-maker	Fartor, ōris, *m.*	3 Ἀλλαντοποιὸς, οῦ, m.
Pudding-meat or Pap	Pappa, æ, *f.*	2 Μάζα, ης, f.
A Black-Pudding or Bloud-Pudding	Botŭlus, i, *m.*	
A Sausage	* Lucanĭca, æ, *f.*	5 Ἀλλᾶς αντος, m.
A Chitterlin	Hilla, æ, *f.*	3 Ἐγκοίλια, ίων, n.
A Flesh-Pie	Artocrea, æ, *f.*	5 c. Ἀρτόκρεας, ατος, n.
An Apple-Pie	Artomēlum, i, *n.*	3 Ἀρτόμηλον, ήλου, n.
A Sallet	Acetarium, ii, *n.*	
Vineger	Acētum, i, *n.*	1 c. Ὄξος, εος ους n.
Oil	Oleum, i, *n.*	3 Ἔλαιον, αίου, n.
Sauce	Intinctus, ûs, *m.*	5 Ἔμβαμμα, ατος, n.
Pickle	Muria, æ, *f.*	2 Ἅλμη, ης, f.
Verjuice of Grapes	Omphacium, ii, *n.*	3 Ὀμφάκιον, ίου, n.
Verjuice of Crabs	Agresta, æ, *f.*	

* Other words there are that signifie a Sausage, as *Isicium* and *Apexabo*; but I made choice of this, because the Italians at this day call a Sausage *Luganica.*

XVI.

Of Drink.	*De Potu.*	Περὶ Πόσεως.
Drink	Potus, ûs, *m.*	2 c. Πόσις, εως, f.
A Draught	Haustus, ûs, *m.*	
Wine	Vinum, i, *n.*	3 Οἶνος, ου, m.
New Wine	Mustum, i, *n.*	1 c. Γλεῦκος, εος ους, n.
Dead Wine	Vappa, æ, *f.*	1 Ἐκτροπίας, ου, m.
White Wine	Vinum Gallicum albū.	Οἶνος Γαλατικὸς λευκὸς
Claret Wine	Vinum Gallic. rubrum	Οἶνος Γαλατικὸς ἐρυθρὸς
Sack	Vinum Hispanicum	
Canary	Vinum Canariense	
Wormwood Wine	Vinum Absinthites	1 Ἀψινθίτης, ου, m.
Mead or Metheglin	Mulsum, i, *n.*	5 Οἰνόμελι, έλιτος, n.

Sider

Sider	Melites, æ, *m.*	1 Μηλίτης, ȣ, m.
Perry	Apites, æ, *m.*	1 Ἀπίτης ȣ, m.
Beer or Ale	Cerevisia, æ, *f.*	3 Ζύθος, ȣ, m.
Beverage	Posca, æ, *f.*	3 Ὀξύκρατον, ȣ, n.
A Brewer	Cerevisiarius, ii, *m.*	3 Ζυθοποιὸς, ȣ̃, m.
Dregs	Fæx, fæcis, *f.*	5 Τρὺξ, γός, f.
Strong-water	Aqua ardens	
A Pot	Pocŭlum, i, *n.*	3 Ποτήριον, ίȣ, n.
A Flagon	Lagēna, æ, *f.*	3 Λάγυνος, ύνȣ, f.
A Beaker	Crater, ēris, *m.*	5 Κρατὴρ, ῆρος, m.
A Bowl	Patĕra, æ, *f.*	
A Jug or Cup	Scyphus, i, *m.*	3 Σκύφος, ȣ, m.
A Bottle or Jack	Uter, tris, *m.*	3 Ἀσκὸς, ȣ̃, m.
A Chalice	Calix, ĭcis, *m.*	5 Κύλιξ, ικος, f.
A Can or Tankard	Canthărus, i, *m.*	3 Κάνθαρος, άρȣ, m.
A Glass	Pocŭlum vitreum, *n.*	3 Ποτήριον ὑάλινον, n.
A Butler	Promus, i, *m.*	1 Ταμίας, ȣ, m.
A Cup-bearer	Pincerna, æ, *m.*	3 Οἰνοχόος, όȣ, m.

XVII.

Of Apparel.	*De Vestitu.*	Περὶ Ἐνδύσεως.
A *Garment* / *Clothing*	Vestis, is, *f.* / Vestitus, ûs, *m.*	3 Ἱμάτιον, ίȣ, n. / 2 c. Ἔσθησις, εως, f.
Cloth	Pannus, i, *m.*	1 c. Ῥάκος, εος ȣς, n.
Linen	Linteum, i, *n.*	2 Ὀθόνη, ης, f.
Canvas	Cannabeum, i, *n.*	3 Καννάβινον, ȣ, n.
Fustian	Xylĭnum, i, *n.*	3 Ξύλινον λίνον, n.
Cotton	Gossipium, ii, *n.*	3 Ἐριόξυλον, ȣ, n.
Silk	Serĭcum, i, *n.*	3 Σηρικὸν, ȣ̃, n.
Velvet	Holoserĭcum, i, *n.*	3 Ὁλοσηρικὸν, ȣ̃, n.
Sack-cloth	Saccus, i, *m.*	3 Σάκκος, ȣ, m.
A Hat	Galērus, i, *m.*	3 Πέτασος, άσȣ, m.

A Hat-

A Hat-band	Spira, æ, *f.*	2 Σπεῖρα, ας, f.
A Cap	Pileus, i, *m.*	3 Πῖλος, ου, m.
A Coif	Capĭtal, ālis, *n.*	5 Κεφαλὶς, ίδος, f.
A Cross-cloth	Plagŭla, æ, *f.*	
A Cawl	Reticŭlum, i, *n.*	3 Κεκρύφαλος ἀλου, m.
A Fillet	Vitta, æ, *f.*	5 Ἄμπυξ, υκος, com.
A Fan	Flabellum, i, *n.*	5 Ῥιπὶς, ίδος, f.
A Bongrace	Umbella, æ, *f.*	3 Σκιάδιον, ίου, n.
A Hood or Veil	Peplum, i, *n.*	3 Πέπλος, ου, m.
An Ear-ring	Inauris, is, *f.*	3 Ἐνώτιον, ου, n.
A Neck-lace	Monīle, is, *n.*	3 Ὅρμος, ου, m.
A Chain	Torquis, is, *com.*	3 Στρεπτὸς, οῦ, m.
A Bracelet	Armilla, æ, *f.*	3 Περιβραχιόνιον, ου, n.
A Spangle	Bractea, æ, *f.*	3 Πέταλον, άλου, n.
A Band	Collāre, is, *n.*	3 Περιτραχήλιον ίου, n.
A Shirt or Smock	Interŭla, æ, *f.* Indusium, (ii, *n.*	3 Χιτωνίσκος, ου, m.
A Wastcoat	Subucŭla, æ, *f.*	5 Ὑποχίτων, ωνος, m.
A Stomacher	Pectorāle, is, *n.*	3 Προστερνίδιον, ου, n.
A Doublet	Diplois, ĭdis, *f.*	5 Διπλοῒς ίδος, f.
A Sleeve	Manĭca, æ, *f.*	5 Χειρὶς, ίδος, f.
A Pair of Breeches	Braccæ, arum, *f.*	5 Ἀναξυρὶς, ιδος, f.
A Button	Fibŭla, æ, *f.*	2 Περόνη, ης, f.
A Point	Ligŭla, æ, *f.*	5 Σφαιρωτὴρ, ῆρος, m.
A Girdle	Cingŭlum, i, *n.*	2 Ζώνη, ης, f.
A Skirt	Fimbria, æ, *f.*	3 Κράσπεδον, ου, n.
A Gown	Toga, æ, *f.*	2 Τήβεννα, ης, f.
A Coat	Tunĭca, æ, *f.*	5 Χιτὼν, ῶνος, m.
A Loose Gown	Stola, æ, *f.*	2 Στόλη, ης, f.
A Jacket	Tunicŭla, æ, *f.*	3 Χιτώνιον, ίου, n.
A Cloak	Pallium, ii, *n.*	1 c. Φάρος εος ους, n Φελόνη
A short Cloak	Chlamys, ydis, *f.*	5 Χλαμὺς, ύδος, f.
A Cassock	Sagum, i, *n.*	3 Σάγος, ου, m.
An Apron	Præcinctorium, ii, *n.*	5 Περίζωμα, ατος, n.
Stockins	Tibialia, ium, *n.*	5 Κνημὶς, ίδος, f.
A Garter	Fascia cruralis	5 Περισκελὶς ίδος, f.

A Boot

A Boot	Ocrea, æ, *f.*	5 Κνημὶς ἴδος, f.
A Spur	Calcar, āris, *n.*	3 Κέντρον, ου, n.
A Shoo	Calceus, i, *m.*	5 Ὑπόδημα, ατος, n.
A Shoo-Latchet	Corrigia, æ, *f.*	5 Ἱμὰς, άντος, m.
A Slipper	Crepĭda, æ, *f.*	5 Κρηπὶς, ἴδος, f.
A Sole	Solea, æ, *f.*	5 Πέλμα, ατος, n.
Socks	Socci, orum, *m.*	1 Ἐμβάται, ων, m.
A Buskin	Cothurnus, i, *m.*	3 Κόθορνος, όρνου, m.
A Glove	Chirothēca, æ, *f.*	2 Χειροθήκη, ης, f.
A Hook or Clasp	Uncus, i, *m.*	3 Ἄγκιστρον, ου, n. Ὄγκος ου.
An Eye	Ocellus, i, *m.*	2 Ὀπὴ, ῆς, f.
A Ring	Annŭlus, i, *m.*	3 Δακτύλιον, ίου, n.
A Veil	Velum, i, *n.*	5 Περικάλυμμα, τος, n.
A Handkerchief	Sudarium, ii, *n.*	3 Ῥινόμακτρον, ου, n.
A Bodkin	Acus crinalis	3 Ξανεῖον, ου, n.
A Pocket	Locŭlus, i, *m.*	
Stilts	Grallæ, arum, *f.*	3 Καλόβαθρα, ων, n.

XVIII.

Of Buildings.	*De Ædificiis & eorum partibus.*	Περὶ Οἰκοδομῶν.
A *Building*	ÆDificium, ii, *n.*	2 Οἰκοδομὴ, ῆς, f.
A City	Urbs, bis, *f.*	2 c. Πόλις, εως, f.
A Gate	Porta, æ, *f.*	2 Πύλη, ης, f.
A Portcullis	Catarracta, æ, *f.*	1 Καταρράκτης, ου, m.
A Tower	Turris, is, *f.*	3 Πύργος, ου, m.
A Castle	Arx, arcis, *f.*	2 c. Ἀκρόπολις, εως, f.
A House	Domus, ûs, *m.*	3 Οἶκος, ου, m.
A Palace	Palatium, ii, *n.*	2 Βασιλικὴ, ῆς, f.
A Cottage	Casa, æ, *f.*	2 Καλύβη, ης, f.
A Conduit	Aquæductus, ûs, *m.*	2 Ὑδραγωγία, ας, f.
A Conduit-Pipe	Tubus, i, *m.*	5 Σύριγξ, γγος, f.

* *An Hospital*	Ptochotrophium, ii, *n.*	3 Πτωχοτροφεῖον, ου, n.
A College	Collegium, ii, *n.*	
A Library	Bibliothēca, æ, *f.*	2 Βιβλιοθήκη, ης, f.
A Court or Sessions-house	Curia, æ, *f.*	3 Ἀρχεῖον, ου, n.
An Exchange	Cambium, ii, *n.*	
A Magazin or Store-house	Repositorium, ii, *n.*	2 Ἀποθήκη, ης, f.
A Senate-house	Senacŭlum, i, *n.*	3 Βουλευτήριον, ίου, n.
A Religious house or Monastery	Monasterium, ii, *n.*	3 Μοναστήριον, ίου, n.
Cloisters	Peristylium, ii, *n.* Porticus, ûs, *f.*	3 Περιστύλιον, ίου, n.
A Play-house	Theātrum, i, *n.*	3 Θέατρον, ἄτρου, n.
A Tennis-Court or Bowling-Alley	Sphæristerium, ii, *n.*	3 Σφαιριστήριον, ίου, n.
A Race	Stadium, ii, *n.*	3 Στάδιον, ίου, n.
The Stand	Carcĕres, um, *m.*	2 Ἀφετηρία, ας, f.
The Goal	Meta, æ, *f.*	5 Καμπτὴρ, ῆρος, m.
An Inn	Diversorium, ii, *n.*	5 Κατάλυμα, ατος, n.
A Tavern	Taberna vinaria, *f.*	3 Οἰνοπώλιον, ίου, n.
A Victualling-house	Caupōna, æ, *f.*	3 Καπηλεῖον, ου, n.
A Cooks Shop	Popīna, æ, *f.*	3 Ὀψοπώλιον, ίου, n.
A Shop or Work-house	Officīna, æ, *f.*	3 Ἐργαστήριον, ίου, n.
The Shambles	Macellum, i, *n.*	3 Κρεωπώλιον, ίου, n.
An Apothecaries Shop	Pharmacopōlium, ii, *n.*	3 Φαρμακοπώλιον ίου, n.
A Jail or Prison	Carcer, ĕris, *n.*	3 Δεσμωτήριον, ίου, n.
A House of Correction	Ergastŭlum, i, *n.*	3 Ἐργαστήριον, ίου, n.
A Pair of Stocks	Cippus, i, *m.*	2 Ποδοκάκη ης, f. Ξύλον.
A Pillory	Numella, æ, *f.*	3 Κλοιὸς, οῦ, m.
A Gallows	Patibŭlum, i, *n.*	3 Σταυρὸς, οῦ, m.
A Market-place	Forum, i, *n.*	2 Ἀγορὰ, ᾶς, f.
A Town	Oppĭdum, i, *n.*	3 Πολίχνιον, ίου, n.
A Street	Platēa, æ, *f.*	2 Πλατεῖα, ας, f.

* Of Hospitals or Alms-houses there are several sorts, either for the Poor only, and then they are called *Ptochotrophea* and *Ptochodochea*; or for the enterteining of Strangers and poor Travellers, and such are called *Xenodochia*; or for the reception and cure of the Diseased, Wounded, *&c.* and those are called *Nosocomia*.

A Lane

A Lane	Vicus, i, *m.*	2 Ῥύμη, ης, f.
An Alley or narrow Lane	Angiportus, ûs, *m.*	3 Στενωπὸς, ȣ, com.
A Countrey Town or Village	Pagus, i, *m.*	2 Κώμη, ης, f.
A Booth	Tentorium, ii, *n.*	2 Σκηνὴ, ης, f.
A Fold	Caula, æ, *f.* } Stabŭlū,	2 Αὐλία, ας, f.
A Stable	Equīle, is, *n.* } i, *n.*	3 Ἱπποστάσιον, ίȣ, n.
A Rack } *A Manger*	Præsēpe, is, *n.*	2 Φάτνη, ης, f.
A Swine-sty	Hara, æ, *f.*	3 Συφεὸς, ȣ, m.
A Coop or Pen	Cors, cortis, *f.*	5 Ὀρνιθὼν, ῶνος, m.
A Pigeon-house	Columbarium, ii, *n.*	5 Περιστερεὼν, ῶνος, m.
A Hall	Aula, æ, *f.*	2 Αὐλὴ, ῆς, f.
A Kitchin	Culīna, æ, *f.*	3 Μαγειρεῖον, ȣ, n.
A Parlour	Conclāve discubitoriū	3 Τρικλίνιον, ίȣ, n.
A Dining-Room	Pransorium, ii, *n.*	
A Bed-chamber	Cubicŭlum, i, *n.*	5 Κοιτὼν, ῶνος, m.
An inner Room (general	Penetrāle, is, *n.* }	3 Μυχὸς, ȣ, m.
A Closet or any Room in	Conclāve, is, *n.* }	
A Study	Musēum, i, *n.*	3 Μȣσεῖον, ȣ, n.
A Privy or House of office	Latrīna, æ, *f.*	5 Ἀφεδρὼν, ῶνος, m.
A Cellar	Cella, æ, *f.*	
A Buttery or Store-house	Promptuarium, ii, *n.*	3 Ταμιεῖον, ȣ, n.
A Larder	Carnarium, ii, *n.*	
A Garret or upper Room	Cœnacŭlum, i, *n.*	3 Ἀνώγεον, ȣ, n.
A Gallery	Pergŭla, æ, *f.*	2 Π[illegible]βολὴ, ῆς, f.
A Bake-house	Pistrīnum, i, *n.*	3 Ἀρτοκοπεῖον, ȣ, n.
A Mill	Mola, æ, *f.*	2 Μύλη, ης, f.
A Brick	Later, ĕris, *m.*	3 Πλίνθος, ȣ, m.
A Tile	Tegŭla, æ, *f.*	3 Κέραμος, άμȣ, m.
A Shingle	Scandŭla, æ, *f.*	5 Σχίδαξ, ακος, f.
Lime	Calx, cis, *f.*	3 Τίτανος, ȣ, f. Κονία, f.
Plaster or Parget	Tectorium, ii, *n.*	5 Κονίαμα, ατος, n.
Mortar	Intrītum, i, *n.*	
Clay	Lutum, i, *n.*	3 Πηλὸς, ȣ, m.
Rubbish	Rudus, ĕris, *n.*	3 Ἐρείπιον, ίȣ, n.

A Foun-

A Foundation	Fundamentum, i, *n.*	3 Θεμέλιον, ίου n.
A Wall	Paries, ĕtis, *m.*	1 c. Τεῖχος, εος-ους n.
A Pillar	Columna, æ, *f.*	5 Κίων, ονος, com.
A Corner	Angŭlus, i, *m.*	2 Γωνία, ας, f.
A Porch	Veſtibŭlum, i, *n.*	3 Πρόθυρον ου, n.
A Penthouſe	Complŭvium, ii, *n.*	3 Κατάκλυστρον, ου, n.
A Gate	Janua, æ, *f.* }	2 Θύρα, ας, f.
A Door	Oſtium, ii, *n.* }	
A Knocker or Ring of a (door	Cornix, īcis, *f.*	2 Κορώνη, ης, f.
A Latch	Clauſtrum, i, *n.*	3 Κλεῖθρον, ου, n.
A Bar	Repagŭlum, i, *n.*	5 Ἐπιβλὴς, ῆτος, m.
A Chink	Rima, æ, *f.*	2 Κλειθρία ας f. Ὀπὴ, ῆς.
A Fore-door	Antīcum, i, *n.*	3 Πρόθυρον ύρου, n.
A Back-door	Poſtīcum, i, *n.*	3 Ψευδόθυρον, ύρου, n.
A Lock	Sera, æ, *f.*	3 Κλεῖθρον, ου, n.
A Key	Clavis, is, *f.*	5 Κλεὶς, ιδὸς, f.
A Bolt	Obex, ĭcis, *d.* Peſsŭlus, i.	3 Μοχλὸς οῦ m.
A Hinge	Cardo, ĭnis, *com.*	3 c. Στρόφιγξ, έως, m.
The Lintel	Superliminăre, is, *n.*	3 Ὑπέρθυρον, ου, n.
A Poſt	Poſtis, is, *m.*	5 Παραστὰς άδος, f.
A Wicket	Oſtiŏlum, i, *n.*	3 Θυρίδιον, ίου, n.
Folding-doors	Valvæ, arum, *f.*	5 Δικλὶς, ίδος, f.
A Threſhold	Limen, ĭnis, *n.*	3 Ὀυδὸς, οῦ, m.
A Window	Feneſtra, æ, *f.*	5 Θυρὶς, ίδος, f.
A Caſement	Tranſenna, æ, *f.*	5 Κιγκλὶς, ίδος, f.
A Lattice	Cancelli, orum, *m.*	3 Δρύφακτος, άκτου, m.
The Roof an houſe	Tectum, i, *n.*	2 Στέγη ης, f.
The Ridge	Faſtigium, ii, *n.*	3 Ἀκρωτήριον ίου, n.
The Eaves	Suggrundium, ii, *n.*	3 Γεῖσον ου, n.
An Arch	Fornix, ĭcis, *m.*	2 Καμάρα, ας f.
A vaulted Roof	Laquear, āris, *n.*	5 Φάτνωμα τος n.
A Prop	Tibīcen, ĭnis, *m.*	5 Ἀντηρὶς, ίδος, f.
A Beam	Trabs, bis, *f.*	3 Δοκὸς, οῦ, f.
A Rafter	Tignum, i, *n.*	5 Δοκὶς, ίδος, f.
A Board	Tabŭla, æ, *f.*	5 Σανὶς, ίδος, f.

A Nail	Clavus, i, *m.*	3 Ἧλος, ου, m.
A Lath	Asserculus, i, *m.*	3 Σανίδιον, ίου, n.
A Pin or Peg	Paxillus, i, *m.*	3 Πάσσαλος, άλου, m.
A Story	Contignatio, ōnis, *f.*	5 Στέγασμα ατος, n.
A Pavement	Pavimentum, i, *n.*	1 c. Ἔδαφος εος ους, n.
The Floor	Solum, i, *n.*	3 Πέδον, ου, n.
Stairs	Scalæ, arum, *f.*	5 Κλίμαξ, ακος, f.
Winding Stairs	Scalæ cochlīdes, *f.*	5 Κλίμαξ ἑλικτή
A Step or Round	Gradus, ûs, *m.*	3 Βαθμὸς, οῦ, m.
A Sieling	Concameratio, ōnis, *f.*	2 c. Καμάρωσις, εως, f.
A Chimney	Camīnus, i, *m.*	2 Καπνοδόχη, ης, f.
A Hearth	Focus, i, *m.*	2 Ἐσχάρα, ας, f.
A Furnace	Fornax, ācis, *f.*	3 Κάμινος, ου, f.
An Oven	Furnus, i, *m.*	3 Ἶπνος, ου, f. Κλίβανος, ου.

XIX.

Of GOD.	*De DEO.*	Περὶ ΘΕΟΥ͂.
GOD	DEus, i, *m.*	3 Θεὸς, οῦ, m.
The Godhead	Deĭtas, ātis, *f.*	5 Θεότης, ητος, f.
Lord	Domĭnus, i, *m.*	3 Κύριος, ίου, m.
Lordship	Dominium, ii, *n.*	5 Κυριότης, ητος, f.
An Attribute	Attribūtum, i, *n.*	3 Ἐπίθετον, ου, n.
Infinite	Infinītus, i, *m.*	3 Ἄπειρος ου, m.
Eternal	Æternus, i, *m.*	3 Ἀΐδιος, ίου, m.
Blessed	Beātus, i, *m.*	3 Μακάριος, ίου, m.
Perfect	Perfectus, i, *m.*	3 Τέλειος, είου, m.
Simple	Simplex, ĭcis, *m.*	3 Ἀσύνθετος, έτου, m.
Immutable	Immutabĭlis, is, *m.*	3 Ἀμετάβολος, όλου, m.
Incomprehensible	Incomprehensibilis, is.	3 Ἀκατάληπτος, ου, m.
Essence	Essentia, æ, *f.*	2 Οὐσία, ας, f.
Unity	Unĭtas, ātis, *f.*	5 Ἑνότης, ότητος, f.
A Person	Persōna, æ, *f.*	2 c. Ὑπόστασις, εως, f.

Trinity	Trinĭtas, ātis, *f.*	5 Τριὰς, ἀδος, f.
The FATHER	Pater, tris, *m.*	5 Πατὴρ, τέρος τρὸς, m.
unbegotten	Ingenĭtus, i, *m.*	3 Ἀγέννητος, ου, m.
The Creator	Creātor, ōris, *m.*	1 Κτίστης, ου, m.
Creation	Creatio, ōnis, *f.*	1 c. Κτίσις, εως, f.
The SON	Filius, ii, *m.*	5 Υἱὸς, οῦ, m.
Begotten	Genĭtus, i, *m.*	3 Γεννητὸς, οῦ, m.
Jesus	Jesus, u, *m.*	Ἰησοῦς, οῦ, m.
Christ	Christus, i, *m.*	3 Χριστὸς, οῦ, m.
A Redeemer	Redemptor, ōris, *m.*	1 Λυτρωτὴς, οῦ, m.
Redemption	Redemptio, ōnis, *f.*	2 c. Λύτρωσις, εως, f.
A Price	Pretium, ii, *n.*	3 Λύτρον, ου, n.
A Saviour	Salvātor, ōris, *m.*	5 Σωτὴρ, ῆρος, m.
Salvation	Salvatio, ōnis, *f.*	2 Σωτηρία, ας, f.
A Mediator	Mediātor, ōris, *m.*	1 Μεσίτης, ου, m.
Incarnation	Incarnatio, ōnis, *f.*	2 c. Ἐνσάρκωσις, εως, f.
Conception	Conceptio, ōnis, *f.*	2 c. Σύλληψις, εως, f.
Nativity	Nativĭtas, ātis, *f.*	2 c. Γέννησις, εως, f.
Temptation	Tentatio, ōnis, *f.*	3 Πειρασμὸς, οῦ, m.
Transfiguration	Transfiguratio, ōnis, *f.*	2 c. Μεταμόρφωσις εως, f.
The Cross	Crux, cis, *f.*	3 Σταυρὸς, οῦ, m.
Crucifixion	Crucifixio, ōnis, *f.*	2 c. Σταύρωσις, εως, f.
A Miracle	Miracŭlum, i, *n.*	5 Θαῦμα, τος, n.
Resurrection	Resurrectio, ōnis, *f.*	2 c. Ἀνάστασις, εως, f.
Ascension	Ascensio, ōnis, *f.*	2 c. Ἀνάβασις, εως, f.
Intercession	Intercessio, ōnis, *f.*	2 c. Ἔντευξις, εως, f.
A Judg	Judex, ĭcis, *com.*	1 Κρίτης, ου, m.
Judgment	Judicium, ii, *n.*	2 c. Κρίσις, εως, f.
Heaven	Cœlum, i, *n.*	3 Οὐρανὸς, οῦ, m.
Glory	Gloria, æ, *f.*	2 Δόξα, ης, f.
Hell	Infernus, i, *m.*	1 Ἅδης, ου, m.
Torment	Tormentum, i, *n.*	3 Βάσανος, άνου, f.
The HOLY GHOST	Spiritus sanctus, *m.*	5 Πνεῦμα ἅγιον, n.
Proceeding	Procedens, tis, *m.*	3 Ἐκπορευόμενος, ου, m.
The Sanctifier	Sanctificātor, ōris, *m.*	5 Ἁγιάζων, οντος, m.

The Comforter	Consolātor, ōris, *m.*	3 Παράκλητος, ήτου, m.
A Gift	Donum, i, *n.*	5 Χάρισμα, τος, n.
Grace	Gratia, æ, *f.*	5 Χάρις, ιτος, f.

XX.

Of created Spirits.	*De Spiritibus creatis.*	Περὶ Πνευμάτων κτιστῶν.
AN Angel	ANgĕlus, i, *m.*	3 Ἄγγελος, έλου, m.
A Spirit	Spirĭtus, ûs, *m.*	5 Πνεῦμα, τος, n.
A Saint	Sanctus, i, *m.*	3 Ἅγιος, ίου, m.
The Divel	Diabŏlus, i, *m.*	3 Διάβολος, όλου, m.
A Fiend	Furia, æ, *f.*	5 Ἀλάστωρ, ορος, m.
A Bugbear	Terriculamentum, i, *n.*	3 Μορμολύκειον, ου, n.
An Apparition	Spectrum, i, *n.*	5 Φάντασμα, ατος, n.
A Witch	Saga, æ, *f.*	2 Φαρμακεύτρια, ίας, f.
A Conjurer	Exorcista, æ, *m.*	1 Ἐξορκιστὴς, οῦ, m.
A Soul	Anĭma, æ, *f.*	2 Ψυχὴ, ῆς, f.
The Vegetative Soul	Anima Vegetativa, æ, *f.*	Αὐξητικὴ Ψυχή
The Sensitive Soul	Anima Sensitiva, æ, *f.*	Αἰσθητικὴ Ψυχή
The Rational Soul	Anima Rationalis, is, *f.*	Λογικὴ Ψυχή
A Man	Vir, vĭri, *m.*	5 Ἀνὴρ, έρος δρός, m.
A Woman	Mulier, ĕris, *f.*	5 Γυνὴ, αικός, f.

XXI. Of

XXI.

Of the Faculties of the Soul.	*De Facultatibus Animæ.*	Περὶ τῶν Δυνάμεων τῆς Ψυχῆς.
	(I.)	
Of the Senses and their Objects.	*De Sensibus & eorum Objectis.*	Περὶ Αἰσθήσεων καὶ Αἰσθητῶν.
THe SIGHT	VIsus, ûs, *m.*	2 c. Ὅρασις, εως, f.
Light	Lux, lūcis, *f.*	5 Φῶς, φωτός, n.
Brightness	Splendor, ōris, *m.*	2 Μαρμαρυγὴ, ῆς, f.
Darkness	Tenĕbræ, arum, *f.*	1 c. Σκότος, εος ους, n.
Shadow	Umbra, æ, *f.*	2 Σκιὰ, ᾶς, f.
Colour	Color, ōris, *m.*	2 Χρόα, ας, f.
White	Albus, a, um	Λευκὸς, ὴ, όν
Black	Niger, a, um	Μέλας, αινα, άν
Gray	Canus, a, um. Cæsius, a, um	Γλαυκὸς, ὴ, όν
Red	Ruber, bra, brum	Ἐρυθρὸς, ὰ, όν
Yellow	Flavus, a, um	Ξανθὸς, ὴ, όν
Blue	Cærulcus, a, um	Κυάνεος, α, ον
Purple	Purpureus, a, um	Πορφύρεος, α, ον
Brown	Fuscus, a, um	Φαιὸς, ὰ, όν
Parti-coloured	Discŏlor, ōris, *m.*	Ἑτερόχροος, όυ, m.
The SMELL	Odorātus, ûs, *m.*	2 c. Ὄσφρησις, εως, f.
A Sent or Smell	Odor, ōris, *m.*	2 Ὀσμὴ, ῆς, f.
A sweet Smell	Fragrantia, æ, *f.*	2 Εὐωδία, ας, f.
A Stink	Fœtor, ōris, *m.*	2 Δυσωδία, ας, f.
The TAST	Gustus, ûs, *m.*	2 c. Γεῦσις, εως, f.
A Relish	Sapor, ōris, *m.*	3 Χυλὸς, ȣ̃, m.
Sweet	Dulcis, e	Γλυκὺς, εῖα, ύ
Bitter	Amārus, a, um	Πικρὸς, ὰ, όν
Sower	Acĭdus, a, um	Ὀξώδης, εος ους, com.

The

The HEARING	Audītus, ûs, *m.*	2 Ἀκοὴ, ῆς, f.
A Sound	Sonus, i, *m.*	3 Ἦχος ȣ, m.
The TOUCH *or Feeling*	Tactus, ûs, *m.*	2 Ἁφὴ ῆς, f.
Hot	Calĭdus, a, um	Θερμὸς, ὴ, όν
Cold	Frigĭdus, a, um	Ψυχρὸς, ὰ, όν
Moist	Humĭdus, a, um	Ὑγρὸς, ὰ, όν
Dry	Siccus, a, um	Ξηρὸς ὰ, όν
Thick	Densus, a, um	Παχὺς, εῖα, ὺ
Thin	Rarus, a, um	Ἀραιὸς, ὰ, όν
Heavy	Gravis, e	Βαρὺς, εῖα, ὺ
Light	Levis, e	Κȣφὸς, ὴ, όν
Hard	Durus, a, um	Σκληρὸς, ὰ, όν
Soft	Mollis, e	Μαλακὸς, ὴ, όν
Tough	Lentus, a, um	Γλισχρὸς, ὰ, όν
Brittle	Fragĭlis, e	Ψαθυρὸς, ὰ, όν

The Substantives of these are,

Heat	Calor, ōris, *m.*	Θερμότης, ητος, f.
Cold	Frigus, ŏris, *n.*	Ψυχρότης, ητος, f.
Moisture	Humidĭtas, ātis, *f.*	Ὑγρότης
Driness	Siccitas	Ξηρότης
Thickness	Densitas	Παχύτης
Thinness	Raritas	Ἀραιότης
Heaviness	Gravitas	Βαρύτης
Lightness	Levitas	Κȣφότης
Hardness	Durities, ei, *f.*	Σκληρότης
Softness	Mollities	Μαλακότης
Toughness	Lentor, ōris, *m.*	Γλισχρότης
Brittleness	Fragilitas	Ψαθυρότης

(II.)

Of the Understanding, Will and Affections.	*De Intellectu, Voluntate & Affectibus.*	Περὶ τȣ̃ Νόȣ, Θελήματος κὴ Παθῶν.
THe MIND	MEns, tis, *f.*	3 Νόος, νȣ̃ς, νόȣ νȣ̃, m
The understanding	Intellectus, ûs, *m*	2 Διάνοια, ας, f.
Reason	Ratio, ōnis, *f.*	3 Λόγος, ȣ, m.
Knowledg	Scientia, æ, *f.*	2 Ἐπιστήμη, ης, f.

Ignorance

Ignorance	Ignorantia, æ, *f.*	2 Ἄγνοια, ας, f.
The Conſcience	Conſcientia, æ, *f.*	2 c. Συνείδησις, εως, f.
Judgment	Judicium, ii, *n.*	2 c. Κρίσις, εως, f.
Counſel	Conſilium, ii, *n.*	2 Βουλή, ῆς, f.
Prudence	Prudentia, æ, *f.*	2 c. Φρόνησις, εως, f.
Indiſcretion	Imprudentia, æ, *f.*	2 Ἀνοησία, ας, f.
Wiſdom	Sapientia, æ, *f.*	2 Σοφία, ας, f.
Folly	Stultitia, æ, *f.*	2 Μωρία, ας, f.
A Fool	Stultus, i, *m.*	3 Μωρός, οῦ, m.
Art	Ars, artis, *f.*	2 Τέχνη, ης, f.
Experience	Experientia, æ, *f.*	2 Πεῖρα, ας, f.
Unskilful	Inexpertus, a, um	3 Ἄπειρος, ου, com.
Faith or Belief	Fides, ei, *f.*	2 c. Πίστις, εως, f.
Opinion	Opinio, ōnis, *f.*	2 Δόξα, ης, f.
Error or Miſtake	Error, ōris, *m.*	2 Πλάνη, ης, f.
Suſpicion	Suſpicio, ōnis, *f.*	2 Ὑποψία, ας, f.
Doubting	Dubitatio, ōnis, *f.*	2 Ἀπορία, ας, f.
Admiration i. *Wondering*	Admiratio, ōnis, *f.*	5 Θαῦμα, ατος, n.
The WILL	Voluntas, ātis, *f.*	5 Θέλημα, ατος, n.
Liberty	Libertas, ātis, *f.*	2 Ἐλευθερία, ας, f.
The AFFECTIONS or Paſſions	Affectus, uum, *m.*	1 c. Πάθος, εος ους, n.
Truſt	Fiducia, æ, *f.*	2 c. Πεποίθησις, εως, f.
Love	Amor, ōris, *m.*	5 Ἔρως, ωτος, m.
Hatred	Odium, ii, *n.*	1 c. Μῖσος, εος ους, n.
Joy	Gaudium, ii, *n.*	2 Χαρά, ᾶς, f.
Grief	Dolor, ōris, *m.*	2 Λύπη, ης, f.
Mirth	Lætitia, æ, *f.*	2 Εὐφροσύνη, ης, f.
Sadneſs	Triſtitia, æ, *f.*	2 Δυσθυμία, ας, f.
Chearfulneſs	Hilaritas, ātis, *f.*	5 Ἱλαρότης, ητος, f.
Deſire	Deſiderium, ii, *m.*	3 Ἵμερος, έρου, m.
Loathing	Fuga, æ, *f.*	2 Φυγή, ῆς, f.
Boldneſs	Audacia, æ, *f.*	1 c. Θράσος, εος ους, n.
Deſpair	Deſperatio, ōnis, *f.*	2 Ἀνελπιστία, ας, f.
Anger	Ira, æ, *f.*	2 Ὀργή, ῆς, f.

Pleaſure

Pleasure	Voluptas, ātis, *f.*	2 Ἡδονὴ, ῆς, f.
Revenge	Vindicta, æ, *f.*	2 c. Ἐκδίκησις, εως, f.
Shame	Pudor, ōris, *m.*	4 c. Αἰδὼς, όος οῦς, f.
Hope	Spes, ei, *f.*	5 Ἐλπὶς, ίδος, f.
Fear	Timor, ōris, *m.*	3 Φόβος, ου, m.
Envy	Invidia, æ, *f.*	3 Φθόνος, ου, m.
Pity	Misericordia, æ, *f.*	1 c. Ἔλεος, εος ους, n.
Scorn or Contempt	Contemptus, ûs, *m.*	2 Ὀλιγωρία, ας, f.
Repentance	Pœnitentia, æ, *f.*	2 Μετάνοια, ας, f.

XXII.

Of Moral Virtues and Vices.	*De Virtutibus Moralibus & Vitiis.*	Περὶ τῶν Ἀρετῶν καὶ Κακιῶν.
Virtue	Virtus, ūtis, *f.*	2 Ἀρετὴ, ῆς, f.
Vice	Vitium, ii, *n.*	2 Κακία, ας, f.
Godly	Pius, a, um	1 c. Εὐσεβὴς, έος οῦς, m.
Godliness	Piĕtas, ātis, *f.*	2 Εὐσέβεια, ας, f.
Ungodliness	Impiĕtas, ātis, *f.*	2 Ἀσέβεια, ας, f.
Honest	Probus, i, *m.*	3 Καλοκάγαθος, ου, m.
Honesty	Probĭtas, ātis, *f.*	2 Καλοκαγαθία ας, f.
Good	Bonus, a, um	3 Χρηστὸς, οῦ, m.
Bad	Malus, a, um	3 Κακὸς, οῦ, m.
Manners	Mores, um, *m.*	1 c. Ἦθος, εος ους, n.
A Custom	Consuetūdo, ĭnis, *f.*	2 Συνήθεια, ας, f.
Prosperity	Res prosperæ	2 Εὐτυχία, ας, f.
Adversity	Res adversæ	2 Δυστυχία ας, f.
A Sin	Peccātum, i, *n.*	5 Ἁμάρτημα, ατος, n.
An Offence	Scandălum, i, *n.*	3 Σκάνδαλον, άλου, n.
An Oversight	Delictum, i, *n.*	5 Παράπτωμα, ατος, n.
Villany or Wickedness	Scelus, ĕris, *n.*	2 Μοχθηρία ας, f. (πλὴξ
A Villain	Furcĭfer, ĕri, *m.*	1 Μαστιγίας ου, m. Νωτ-
Happiness	Felicĭtas, ātis, *f.*	2 Εὐδαιμονία, ας, f.

Misery

Misery	Miseria, æ, *f.*	5 Ἀθλιότης, ητος, f.
Reward	Præmium, ii, *n.*	3 Μισθὸς, ȣ͂, m.
Punishment	Pœna, æ, *f.*	2 Τιμωρεία, ας, f.
Temperance	Temperantia, æ, *f.*	2 Σωφροσύνη, ης, f.
Pleasure	Voluptas, ātis, *f.*	2 Ἡδονὴ, ῆς, f.
Sobriety	Sobriĕtas, ātis, *f.*	5 Νηφαλιότης, ητος, f.
A Glutton	Heluo, ōnis, *m.*	3 Γαστρίμαργος, ȣ, m
Gluttony	Ingluvies, ei, *f.*	2 Γαστριμαργία, ας, f.
Drunkenness	Ebriĕtas, ātis, *f.*	2 Μέθη, ης, f.
Drunken	Ebrius, a, um	3 Μέθυσος, ȣ, m.
A Drunkard	Ebriōsus, a, um	5 Οἰνόφλυξ, υγος, m.
A Good-fellow	Combĭbo, ōnis, *m.*	1 Συμπότης, ȣ, m.
A Sluggard	Somniculōsus, i, *m.*	1 c. Ὑπνώδης, εος, ȣς, c.
Cleanliness	Mundities, ei, *f.*	5 Καθαριότης, ητος, f.
Chastity	Castĭtas, ātis, *f.*	2 Ἁγνεία, ας, f.
Lust	Libīdo, ĭnis, *f.*	2 Ἐπιθυμία, ας, f.
Wantonness	Lascivia, æ, *f.*	2 Ἀσέλγεια, ας, f.
Adultery	Adulterium, ii, *n.*	2 Μοιχεία, ας, f.
An Adulterer	Adulter, ĕri, *m.*	3 Μοιχὸς, ȣ͂, m.
Fornication	Fornicatio, ōnis, *f.*	2 Πορνεία, ας, f.
A Baud	Leno, ōnis, *m.*	3 Πορνοβοσκὸς, ȣ͂, m.
A Whore	Merĕtrix, īcis, *f.*	2 Πόρνη, ης, f.
A Harlot	Pellex, ĭcis, *f.*	2 Παλλακὴ, ῆς, f.
Riches	Divitiæ, arum, *f.*	3 Πλȣ͂τος, ȣ, m.
Poverty	Paupertas, ātis, *f.*	2 Πενία, ας, f.
Want	Inopia, æ, *f.*	2 Ἀπορεία, ας, Ἔνδεια, f.
Magnificence	Magnificentia, æ, *f.*	2 Μεγαλοπρέπεια, ας, f.
Magnificent	Magnificus, i, *m.*	1 c. Μεγαλοπρεπὴς, έος, ȣ͂ς.
Liberality	Liberalĭtas, ātis, *f.*	5 Ἐλευθεριότης, ητος, f.
Bounty	Benignĭtas, ātis, *f.*	5 Χρηστότης, ητος, f.
Alms	Eleemosyna, æ, *f.*	2 Ἐλεημοσύνη, ης, f.
A Gift	Donum, i, *n.*	3 Δῶρον, ȣ, n.
A Present	Munus, ĕris, *n.*	5 c. Γέρας, ατος, αος, ως, n.
Thrifty	Frugi, *indec.*	5 Σώφρων, ονος, com.
Frugality or Thrift	Parsimonia, æ, *f.*	4 c. Φειδὼ, όος, ȣ͂ς, f.

Charity	Charĭtas, ātis, *f.*	2 Ἀγάπη, ης, f.
A Good-turn	Beneficium, ii, *n.*	5 Εὐεργέτημα, ατος, n.
An Ill-turn	Maleficium, ii, *n.*	5 Κακούργημα, ατος, n.
Riotousness	Luxuria, æ, *f.*	2 Τρυφὴ, ῆς, f.
Prodigality	Prodigalĭtas, ātis, *f.*	2 Ἀσωτία, ας, f.
Covetousness	Avaritia, æ, *f.*	2 Φιλαργυρία, ας, f.
Covetous	Avārus, i, *m.*	3 Φιλάργυρος, ύρου, m.
Diligence	Diligentia, æ, *f.*	2 Ἐπιμέλεια, ας, f.
Diligent	Diligens, tis, *m.*	1 c. Ἐπιμελὴς, έος οῦς, c.
Negligence	Negligentia, æ, *f.*	2 Ἀμέλεια, ας, f. Ῥαθυμία
Idle	Otiōsus, a, um	3 Ἀργὸς, οῦ, m.
Idleness	Ignavia, æ, *f.*	2 Ἀργία, ας, f.
Sloth	Pigritia, æ, *f.*	2 Ὀκνεία, ας, f.
Honour	Honor, ōris, *m.*	2 Τιμὴ, ῆς, f.
Glory	Gloria, æ, *f.*	2 Δόξα, ης, f.
Praise	Laus, laudis, *f.*	3 Ἔπαινος, αίνου, m.
Dispraise	Vituperium, ii, *n.*	3 Ψόγος, ου, m.
Fame	Fama, æ, *f.*	2 Φήμη, ης, f.
Credit	Existimatio, ōnis, *f.*	2 Εὐδοξία, ας, f.
Disgrace	Dedĕcus, ŏris, *n.*	2 Αἰσχύνη, ῆς, f.
Reproach	Ignominia, æ, *f.*	2 Ἀτιμία, ας, f.
Magnanimity	Magnanimĭtas, ātis, *f*	2 Μεγαλοψυχία, ας, f.
Magnanimous	Magnanĭmus, i, *m.*	3 Μεγαλόψυχος, ύχου, m.
Ambition	Ambitio, ōnis, *f.*	2 Φιλοδοξία, ας, f.
Haughtiness	Arrogantia, æ, *f.*	2 Ὑπερηφανία, ας, f.
Modesty	Modestia, æ, *f.*	5 Κοσμιότης, ητος, f.
Boasting	Jactantia, æ, *f.*	2 Ἀλαζονεία, ας, f.
Impudence	Impudentia, æ, *f.*	2 Ἀναίδεια, ας, f.
Bashfulness	Verecundia, æ, *f.*	4 c. Αἰδὼς, όος οῦς, f.
Bashful	Verecundus, a, um	5 Αἰδήμων, ονος, com.
Sawcy	Procax, ācis, *m.*	3 Ἀναίσχυντος, ου, m.
Sawciness	Procacĭtas, ātis, *f.*	2 Ἀναισχυντία, ας, f.
Humble	Humĭlis, is, *m.*	5 Ταπεινόφρων, ονος, c.
Humility	Humilĭtas, ātis, *f.*	2 Ταπεινοφροσύνη, ης, f.
Proud	Superbus, i, *m.*	3 Ὑπερήφανος, άνου, m.

Pride

Pride	Superbia, æ, *f.*	2 Ὑπερηφανία, ας, f.
Valiant	Fortis, is, *m.*	3 Ἀνδρεῖος, ου, m.
Valour	Fortitūdo, ĭnis, *f.*	2 Ἀνδρεία, ας, f.
Courage	Animi, ōrum, *m.*	3 Θύμος, ου, m.
Boldness	Audacia, æ, *f.*	1 c. Θράσος, εος ους, n.
Rashness	Temerĭtas, ātis, *f.*	2 Προπέτεια, είας, f.
Rash	Temerarius, ii, *m.*	1 c. Προπετὴς, έος οῦς, c.
Patience	Patientĭa, æ, *f.*	2 Ὑπομονὴ, ῆς, f.
Patient	Patiens, tis, *m.*	3 Ὑπομονητικὸς, οῦ, m.
Cowardise	Pusillanimĭtas, ātis, *f.*	2 Μικροψυχία, ας, f.
A Coward	Pusillanĭmus, i, *m.*	3 Μικρόψυχος, ου, m.
Justice	Justitia, æ, *f.*	2 Δικαιοσύνη, ης, f.
Just	Justus, i, *m.*	3 Δίκαιος, αίου, m.
A Knave	Nebŭlo, ōnis, *m.*	3 Πανοῦργος, ου, m.
A Thief	Fur, furis, *com.*	1 Κλέπτης, ου, m.
A Robber	Latro, ōnis, *m.*	1 Ληστὴς, οῦ, m.
A Church-Robber	Sacrilēgus, i, *m.*	3 Ἱερόσυλος, ου, m.
A Cut-purse	Crumenisēca, æ, *m.*	3 Βαλαντιοτόμος, ου, m.
A Rogue	Vagabundus, i, *m.*	1 Ἀλήτης, ου, m.
Equity	Æquĭtas, ātis, *f.*	2 Ἐπιείκεια, ας, f.
Deceit	Fraus, fraudis, *f.*	2 Ἀπάτη, ης, f.
Guile	Dolus, i, *m.*	3 Δόλος, ου, m.
Craft	Astutia, æ, *f.*	2 Πανουργία, ας, f.
Mildness	Mansuetūdo, ĭnis, *f.*	5 Πρᾳότης, ητος, f.
Meekness	Clementia, æ, *f.*	2 Ἐπιείκεια ας, f. (κτος
Mercy	Misericordia, æ, *f.*	2 Ἐλεημοσύνη, ης, f. Οἶ-
Pitiful	Misericors, dis, *m.*	5 Οἰκτίρμων, ονος, m.
Pardon	Venia, æ, *f.*	2 Συγγνώμη, ης, f.
Forgiveness	Remissio, ōnis, *f.*	2 c. Ἄφεσις, εως, f.
Kindness	Benevolentia, æ, *f.*	2 Εὐμένεια, ας, m.
Cruelty	Sævitia, æ, *f.*	5 Ὠμότης, ητος, f.
Strife	Lis, lĭtis, *f.*	5 Ἔρις, ιδος, f.
Quarrelling	Contentio, ōnis, *f.*	2 Φιλονεικία, ας, f.
Chiding	Jurgium, ii, *n.*	1 c. Νεῖκος, εος ους, n.
Reproof	Reprehensio, ōnis, *f.*	2 Ἐπιτιμία, ας, f.

Truth	Veritas, ātis, *f.*	2 Ἀλήθεια, είας, f.
True	Verus, a, um	1 c. Ἀληθὴς, έος οῦς, c.
Soothing	Obſequium, ii, *n.*	2 Κολακεία, ας, f.
A Flatterer	Adulātor, ōris, *m.*	5 Κόλαξ, ακος, m.
A Lier	Mendax, ācis, *m.*	1 Ψεύστης, ου, m.
A Lie	Mendacium, ii, *n.*	1 c. Ψεῦδος, εος ους, n.
A Prater	Garrŭlus, i, *m.*	3 Ἀδόλεσχος, ου, m.
Babling or Pratling	Garrulĭtas, ātis, *f.*	2 Ἀδολεσχία, ας, f.
A Buſie-body	Ardelio, ōnis, *m.*	5 Πολυπράγμων ονος, m
A Trifler	Nugātor, ōris, *m.*	3 Φλύαρος, ου, m.
Trifles	Nugæ, arum, *f.*	2 Φλυαρία, ας, f.
Obedient	Obediens, tis, *m.*	1 c. Εὐπειθὴς, έος οῦς, c.
Rebellious	Rebellis, is, *m.*	1 c. Ἀπειθὴς, έος οῦς, c.
Stubborn	Contŭmax, ācis, *m.*	1 c. Αὐθάδης, εος ους, c.
Stubbornneſs	Contumacia, æ, *f.*	2 Αὐθάδεια, ας, f.
Thankful	Gratus, i, *m.*	3 Εὐχάριστος, ίςου, m.
Thankfulneſs	Gratitūdo, ĭnis, *f.*	2 Εὐχαριστία, ας, f.
Unthankful	Ingrātus, i, *m.*	3 Ἀχάριστος, ίςου, m.
Unthankfulneſs	Ingratitūdo, ĭnis, *f.*	2 Ἀχαριστία, ας, f.
Civility	Urbanĭtas, ātis, *f.*	2 Εὐτραπελία, ας, f.
Clowniſhneſs	Ruſticĭtas, ātis, *f.*	2 Ἀγροικία, ας, f.
Courteous	Comes, ĭtis, *m.*	5 Φιλόφρων, ονος, m.
Courteſie	Comĭtas, ātis, *f.*	2 Φιλοφροσύνη, ης, f.
A Jeſt	Jocus, i, *m.*	2 Παιδιὰ, ᾶς, f.
A Scoff	Scomma, ătis, *n.*	5 Σκῶμμα, ατος, n.
A Reproach	Opprobrium, ii, *n.*	1 c. Ὄνειδος, εος ους, n.
A Mocking-Stock	Ludibrium, ii, *n.*	3 Καταγέλαστος, ου, com.
Friendſhip	Amicitia, æ, *f.*	2 Φιλία, ας, f.
Enmity	Inimicitia, æ, *f.*	2 Ἔχθρα, ας, f.
Concord	Concordia, æ, *f.*	2 Ὁμόνοια, ας, f.
Diſcord	Diſcordia, æ, *f.*	2 Διαφωνία, ας, f.
A Grudge	Simultas, ātis, *f.*	
Peace	Pax, pacis, *f.*	2 Εἰρήνη, ης, f.

XXIII. Of

XXIII.

Of Kindred and Affinity.	*De Cognatione & Affinitate.*	Περὶ Συγγενείας καὶ Ἀγχιστείας.
A Genealogy	Genealogia, æ, *f.*	2 Γενεαλογία, ας, f.
Anceſtors	Majōres, um, *m.*	3 Πρόγονοι, όνων, m.
Poſterity	Minōres, um, *m.*	3 Ἀπόγονοι, όνων, m.
Kindred by the Father	Agnatio, ōnis, *f.*	2 Συγγένεια, ας, f.
Kindred by the Mother	Cognatio, ōnis, *f.*	
A Father	Pater, tris, *m.*	5 Πατὴρ, τέρος τρός, m.
A Mother	Mater, tris, *f.*	5 Μήτηρ, τέρος τρός, f.
A Grandfather	Avus, i, *m.*	3 Πάππος, ου, m.
A Grandmother	Avia, æ, *f.*	2 Τήθη, ης, f.

The ſuperior Degrees in a right Line aſcending are, 1. *Proavus*, 2. *Abavus*, 3. *Atavus*, 4. *Tritavus*.

Children	Libĕri, orum, *m.*	3 Τέκνα, ων, n.
A Son	Filius, ii, *m.*	3 Υἱὸς, οῦ, m.
A Daughter	Filia, æ, *f.*	5 Θυγάτηρ, τέρος τρός, f.
A Grandchild { *Male*	Nepos, ōtis, *m.*	3 Υἱωνὸς, οῦ, m.
A Grandchild { *Female*	Neptis, is, *f.*	2 Υἱώνη, ης, f.
A great Grandchild	Pronĕpos, ōtis, *m.*	3 Προέκγονος, ου, m.
A Brother	Frater, tris, *m.*	3 Ἀδελφὸς, οῦ, m.
A Siſter	Soror, ōris, *f.*	2 Ἀδελφὴ, ῆς, f.
Twins	Gemelli, orum, *m.*	3 Δίδυμοι, ύμων, m.
An Uncle by the { *Father*	Patruus, i, *m.*	3 Πατράδελφος, ου, m.
An Uncle by the { *Mother*	Avuncŭlus, i, *m.*	3 Μητράδελφος, ου, m.
An Aunt by the { *Father*	Amĭta, æ, *f.*	2 Πατραδέλφη, ης, f.
An Aunt by the { *Mother*	Matertĕra, æ, *f.*	2 Μητραδέλφη, ης, f.
A Coſin German by the { *Father he & ſhe*	Patruēlis, is, *com.*	3 Ἀνεψιὸς, οῦ, m.
A Coſin German by the { *Mother* { *he*	Conſobrīnus, i, *m.*	
A Coſin German by the { *Mother* { *ſhe*	Conſobrīna, æ, *f.*	2 Ἀνεψιὰ, ᾶς, f.
Affinity	Affinĭtas, ātis, *f.*	2 Ἀγχιστεία, ας, f.
A Kinſman or woman by marriage	Affinis, is, *com.*	3 c. Ἀγχιστεὺς, έως, com.

Wedlock

Wedlock	Conjugium, ii, *n.*	2 Συζυγία, ας, f.
Marriage	Nuptiæ, arum, *f.*	3 Γάμος, ου, m.
Espousals	Sponſalia, orum, *n.*	3 Μνηστεύματα, άτων, n.
* *An Husband*	Marītus, i, *m.*	1 Ἀκοίτης, ου, m.
* *A Wife*	Uxor, ōris, *f.*	2 c. Ἄκοιτις, ιος, f.
A Single Perſon	Cœlebs, ĭbis, *m.*	3 Ἄγαμος, ου, m.
A Virgin	Virgo, ĭnis, *f.*	3 Παρθένος, ου, f.
A Bridegroom	Sponſus, i, *m.*	3 Νυμφίος, ίου, m.
A Bride	Sponſa, æ, *f.*	2 Νύμφη, ης, f.
A Brideman	Pronŭbus, i, *m.*	5 Προμνηστὴρ, ῆρος, m.
A Bridemaid	Pronŭba, æ, *f.*	2 Προμνήστρια, ας, f.
A Bridechamber	Thalămus, i, *m.*	3 Θάλαμος, άμου, m.
A Wooer	Procus, i, *m.*	5 Μνηστὴρ, ῆρος, m.
A Widower	Viduus, i, *m.*	3 Χῆρος, ου, m.
A Widow	Vidua, æ, *f.*	2 Χήρα, ας, f.
A Father in Law	Socer, ĕri, *m.*	3 Πενθερὸς, οῦ, m.
A Mother in Law	Socrus, ûs, *f.*	2 Πενθερὰ, ᾶς, f.
A Son in Law	Gener, ĕri, *m.*	3 Γαμβρὸς, οῦ, m.
A Daughter in Law	Nurus, ûs, *f.*	3 Νυὸς, οῦ, f.
A Brother in Law	Levir, ĭri, *m.*	5 Δαὴρ, έρος, m.
A Siſter in Law	Glos, ōris, *f.*	5 Γάλως, ωος, f.
A Step-Father	Vitrĭcus, i, *m.*	3 Πατρωὸς, οῦ, m.
A Step-Son	Privignus, i, *m.*	3 Πρόγονος, ου, m.
A Step-Mother	Noverca, æ, *f.*	2 Μητρυιὰ, ᾶς, f.
A Step-Daughter	Privigna, æ, *f.*	2 Προγόνη, ης, f.
A Brothers Wife	Fratria, æ, *f.*	5 Ἐινάτηρ, ερος, f.
†*A Nephew*	Nepos ex fratre aut ſorore	3 Ἀδελφοῦ ἢ ἀδελφῆς υἱός
†*A Niece*	Neptis ex fratre aut ſorore	5 Ἀδελφοῦ ἢ ἀδελφῆς θυγάτηρ

* The words Ἀκοίτης and Ἄκοιτις for Husband and Wife are chiefly, if not only uſed by Poets, for which Proſe-writers uſe only Ἀνὴρ and Γυνὴ, ſignifying ſimply Man and Woman.

†The words Nephew and Niece are by us now almoſt wholly reſtrain'd to the Brothers or Siſters Son and Daughter, notwithſtanding *Nepos* and *Neptis*, of which they are derived, ſignifie them we call Grand-children male and female. The Antients never uſed *Nepos* and *Neptis* for the Brothers or Siſters Son and Daughter, whom they called *Fratris aut Sororis filius & filia.*

An Heir	Hæres, ēdis, *m.*	3 Κληρονόμος, ου, m.
A Joint or Co-Heir	Cohæres, ēdis, *m.*	3 Συγκληρονόμος, ου, m.
An Inheritance	Hærēdĭtas, ātis, *f.*	2 Κληρονομία, ας, f.
An Orphan	Orphănus, i, *m.*	3 Ὄρφανος, άνου, m.
A Baſtard	Spurius, ii, *m.*	3 Νόθος, ου, m.
A Maſter	Herus, i, *m.*	3 Κύριος, ίου, m.
A Dame	Hera, æ, *f.*	3 Δέσποινα, νης, f.
A Servant	Famŭlus, i, *m.*	3 Δοῦλος, ου, m.
The Maſter of the Houſe	Paterfamilias, *m.*	1 Οἰκοδεσπότης, ου, m.
The Miſtreſs of the Houſe	Materfamilias, *f.*	2 Οἰκοδέσποινα, ης, f.
A Woman in Childbed	Puerpĕra, æ, *f.*	4 c. Λεχὼ, όος οῦς, f.
A Midwife	Obſtĕtrix, īcis, *f.*	2 Μαῖα, ας, f.
A Nurſe	Nutrix, īcis, *f.*	3 Τροφὸς, οῦ, f.
An Hoſt	Hoſpes, ĭtis, *com.*	3 Ξενοδόχος, ου, m.
A Gueſt	Hoſpes, ĭtis, *com.*	3 Ξένος, ου, m.

XXIV.

Of Houſholdſtuff.	*De Supellectile.*	Περὶ Κατασκευῆς.
Utenſils	Utenſilia, ium, *n.*	3 Χρηστήρια, ίων, n.
Furniture	Apparātus, ûs, *m.*	3 Ἔπιπλα, ων, n.
A Cupbord or Side-table	Abăcus, ci, *m.*	5 Ἄβαξ, ακος, m.
A Seat	Sedes, is, *f.*	3 Θάκος, ου, m.
A Chair	Cathĕdra, æ, *f.*	2 Καθέδρα, ας, f.
A Stool	Sella, æ, *f.*	2 Ἕδρα, ας, f.
A Bench	Scamnum, i, *n.*	3 Θρᾶνος, ου, m.
A Foot-ſtool	Scabellum, i, *n.*	3 Ὑποπόδιον, ίου, n.
A Cuſhion	Pulvīnus, i, *m.*	3 Προσκεφάλαιον, αίου, n.
A Carpet	Tapes, ētis, *m.*	5 Τάπης, ητος, m.
A great Cheſt or Ark	Arca, æ, *f.*	3 Κιβωτὸς, οῦ, m.
A Cheſt, Coffer or Cupbord	Ciſta, æ, *f.*	2 Κίστη, ης, f.
A Box	Pyxis, ĭdis, *f.*	5 Πύξις, ιδος, f.
A Trunk	Riſcus, ci, *m.*	3 Ῥίσκος, ου, m.
A Cabinet	Capsŭla, æ, *f.*	3 Κιβώτιον, ίου, n.
A Preſs for Clothes	Veſtiarium, ii, *n.*	3 Ἱματιοφυλάκιον, ίου, n.

A

A Case	Theca, æ, *f.*	2 Θήκη, ης, f.
A Basket	Corbis, is, *dub.*	3 Κόφινος, ίνε, m.
An Hand-basket	Calăthus, i, *m.*	3 Κάλαθος, άθε, m.
A Frail	Sporta, æ, *f.*	5 Σπυρὶς, ίδος, f.
A Panier or Bread-basket	Caniſtrum, i, *n.*	5 Κάνης, ητος, m.
A Flasket	Qualus, i, *m.*	3 Τάλαρος, άρε, m.
A Wallet	Mantĭca, cæ, *f.*	3 Βαλάντιον, ίε, n.
A Veſſel	Vas, vaſis, *n.*	3 Ἀγγεῖον, ε, n.
A Waſhing-tub	Labrum, i, *n.*	3 Πλυνὸς, ȣ, m.
A But or Pipe	Culeus, i, *m.*	
A Barrel	Dolium, ii, *n.*	3 Πίθος, ε, m.
A Kilderkin	Cadus, i, *m.*	3 Κάδος, ε, m.
A Rundlet	Doliŏlum, i, *n.*	2 Πιθάκνη, ης, f.
A Bung-hole	Orificium, ii, *n.*	5 Στόμα, ατος, n.
A Tap, Spigot or Stopcock	Epiſtomium, ii, *n.*	3 Ἐπιστόμιον, ίε, n.
A Faucet	Fiſtŭla, æ, *f.*	5 Σύριγξ, ιγγος, f.
A Cock	Siphuncŭlus, i, *m.*	5 Σίφων, ωνος, m.
A Funnel	Infundibŭlum, i, *n.*	2 Χοάνη, ης, f.
A Bucket or Pail	Sitŭla, æ, *f.*	5 Κάλπις, ιδος, f.
A Milk-pail	Mulctrāle, is, *n.*	2 Πέλλα, ης, f. Γαυλὸς, ȣ.
A Strainer or Colander	Colum, i, *n.*	3 Ἠθμὸς, ȣ, m.
A Milk-bowl	Sinum, i, *n.*	5 Σκαφὶς, ίδος, f.
A Pump	Antlium, ii, *n.*	3 Ἄντλιον, ίε, n.
A Ciſtern	Ciſterna, æ, *f.*	2 Δεξαμενὴ, ῆς, f.
A Lid or Cover	Opercŭlum, i, *n.*	5 Πῶμα, ατος, n.
An Handle	Anſa, æ, *f.*	2 Λαβὴ, ῆς, f.
A Rack or Cob-Iron	Crateuterium, ii, *n.*	3 Κρατευτήριον, ίε, n.
A Spit	Veru, *n. indec.*	3 Ὀβελος, έλε, m.
Dripping	Liquāmen, ĭnis, *n.*	5 Τῆγμα, ατος, n.
A Dreſſer	Menſa coquinaria	3 Ἐλεὸς, ȣ, m.
A Caldron	Lebes, ētis, *m.*	5 Λέβης, ητος, m.
A Fleſh-hook	Fuſcinŭla, æ, *f.*	2 Κρεάγρα, ας, f.
A Kettle or Pan	Cacăbus, i, *m.*	2 Κακκάβη, ης, f.
A Poſnet or Skillet	Auxilla, æ, *f.*	
A Pipkin	Ollŭla, æ, *f.*	5 Χυτρόπους, οδος, m.

A

A Pot	Olla, æ, *f.*	2 Χύτρα, ας, f.
A Trivet	Tripos, ŏdis, *m.*	5 Τρίπους, οδος, m.
A Chafing-dish	Focŭlus, i, *m.*	3 Πύραυνον, άνου, n.
A Frying-Pan	Sartāgo, ĭnis, *f.*	3 Τήγανον, άνου, n.
A Trey	Trulla, æ, *f.*	
A Ladle	Cochleāre majus, *n.* Tudicŭla, æ, *f.*	2 Τορύνη, ης, f.
A Grater	Radŭla, æ, *f.*	2 c. Κνῆστις, εως, f.
A Pitcher	Urceus, i, *m.*	3 Κεράμιον, ίου, n.
A Gridiron	Craticŭla, æ, *f.*	5 Ἐσχαρὶς, ίδος, f.
A Mortar	Mortarium, ii, *n.*	3 Ὅλμος, ου, m.
A Pestle	Pistillum, i, *n.*	5 Δοίδυξ, υκος, m.
A Dish-clout or Mop	Penicŭlus, i, *m.*	
A Besom	Scopa, æ, *f.*	3 Σάρωθρον, ώθρου, n.
A Kneading-Trough	Mactra, æ, *f.*	2 Μάκτρα, ας, f.
A Fire-Shovel } *A Warming-Pan* }	Batillus, i, *m.*	3 Πύραυνον, άνου, n.
A Pair of Tongs	Forceps, ĭpis, *com.*	2 Πυράγρα, ας, f.
A Pair of Bellows	Follis, is, *m.*	2 Φῦσα, ης, f.
A Match	Sulphurātum, i, *n.*	
Tinder	Fomes, ĭtis, *m.*	3 Πυρεῖον, ου, n.
A Candle	Lucerna, æ, *f.*	3 Λύχνος, ου, m.
A Candle-week	Ellychnium, ii, *n.*	3 Ἐλλύχνιον, ίου, n.
A Candle-stick	Candelābrum, i, *n.*	3 Λυχνεῖον, ου, n.
A Lanthorn	Laterna, æ, *f.*	3 Λυχνοῦχος, ου, m.
A Torch	Fax, facis, *f.* Tæda, æ, *f.* }	5 Δᾲς, ᾳδός, f.
A Link	Funale, is, *n.* }	
A Lamp	Lampas, ădis, *f.*	5 Λαμπὰς, άδος, f.
A Wax-taper	Cereus, i, *m.*	5 Κηρίων, ίωνος, m.
A Snuff	Fungus, i, *m.*	5 Μύκης, ητος, m.
A Pair of Snuffers	Emunctorium, ii, *n.*	3 Ἀπομυκτήριον, ίου, n.
An Extinguisher	Extinctorium, ii, *n.*	3 Σβεστήριον, ίου, n.
A Bed	Lectus, i, *m.*	3 Λέκτρον, ου, n.
A Bedstead	Sponda, æ, *f.*	3 Τοῖχος, ου, m.
A Curtain	Cortīna, æ, *f.*	5 Περιπέτασμα, ατος, n.

Hangings	Aulæa, orum, *n.*	5 Περιστρώματα, άτων, n.
A Coverlet	Periſtrōma, ătis, *n.* }	5 Περίστρωμα, ατος, n.
A Blanket	Stragŭlum, i, *n.* }	
A Rug	Gausăpe, is, *n.*	5 Ἀμφιτάπης, ητος, m.
A Sheet	Lodix, īcis, *f.*	2 Τύλη, ης, f.
A Mat	Teges, ĕtis, *f.*	3 Ψίαθος, άθου, m.
A Bed or Quilt	Stratum, i, *n.*	2 Στρώμνη, ης, f.
A Bed-tike	Culcĭtra, æ, *f.*	2 Τύλη, ης, f.
A Pillow	Pulvīnar, āris, *n.*	3 Προσκεφάλαιον, αίου, n.
A Bolster	Cervīcal, ālis, *n.*	3 Ὑπαυχένιον, ίου, n.
A Canopy	Conopēum, i, *n.*	3 Κωνωπεῖον, ου, n.
A Pallet-Bed	Grabbātus, i, *m.*	3 Κράββατος, άτου, m.
A Cradle	Cunæ, arum, *f.*	
A Rattle	Crepundia, orum, *n.*	3 Κροτοπαίγνια ίων, n.
A Chamber-Pot	Matŭla, æ, *f.*	5 Ἀμὶς, ίδος, f.
A Close-stool	Sella familiaris	3 Λάσανον, άνου, n.
A Comb	Pecten, ĭnis, *m.*	5 Κτεὶς, ενός, m.
A Brush	Scopŭla, æ, *f.*	
A Curling-Iron	Calamiſtrum, i, *n.*	5 Καλαμὶς, ίδος, f.
A Looking-Glass	Specŭlum, i, *n.*	3 Ἔσοπτρον, ου, n.
Spectacles	Conſpicillum, i, *n.*	
A Thimble	Digitāle, is, *n.*	2 Δακτυλήθρα, ας, f.
A Needle	Acus, ûs, *f.*	5 Ῥαφὶς, ίδος, f.
A Needles-eye	Forāmen acûs	2 Τρυμαλιὰ, ᾶς, f.
A Pin	Acicŭla, æ, *f.*	3 Ἀκέστειον, ίου, n. Βελόνη.
A Distaff	Colus, i, *f.*	2 Ἠλακάτη, ης, f.
A Spindle	Fuſus, i, *m.*	3 Ἄτρακτος, ου, com.
A Wherl	Verticillus, i, *m.*	3 Σφόνδυλος, ύλου, m.
A Reel	Rhombus, i, *m.*	3 Ῥόμβος, ου, m.
A pair of Sheers or Scissers	Forfex, ĭcis, *f.*	5 Ψαλὶς, ίδος, f.
A Purse	Crumēna, æ, *f.*	3 Βαλάντιον, ίου, n.
A Bottom of Yarn	Glomus, i, *m.*	
A Thread	Filum, i, *n.*	3 Μίτος, ου, m.
A Tooth-pick	Dentiſcalpium, ii, *n.*	5 Ὀδοντογλυφὶς ίδος, f.
An Ear-pick	Auriſcalpium, ii, *n.*	5 Ὠτογλυφὶς, ίδος, f.

XXV. Of

XXV.

Of a School.	*De Schola.*	Περὶ Σχολῆς.
A School-master	Ludimagister, tri, *m*	3 Γραμματοδιδάσκαλος (ου m.
An Usher	Hypodidascălus, i, *m*	3 Ὑποδιδάσκαλος ου m
A Master	Præceptor, ōris, *m.* Magister	1 Παιδευτὴς, οῦ, m. Παιδαγωγὸς
A Scholar	Discipŭlus, i, *m.*	1 Μαθητὴς οῦ, m.
A School-fellow	Condiscipŭlus, i, *m.*	1 Συμμαθητὴς οῦ, m.
The Masters Chair	Cathĕdra, æ, *f.*	2 Καθέδρα, ας, f.
A Form	Classis, is, *f.*	2 c. Τάξις, εως, f.
A Seat	Subsellium, ii, *n.*	3 Βάθρον, ου, n.
A Desk or a Press for books	Pluteus, i, *m.*	
Teaching	Institutio, ōnis, *f.*	2 c. Παίδευσις, εως, f.
Learning	Disciplīna, æ, *f.*	2 c. Μάθησις, εως, f.
A Book	Liber, bri, *m.*	3 Βίβλος, ου, f.
Writing-Tables	Pugillāres, ium, *m.*	3 Δέλτοι, ων, f.
A Leaf	Folium, ii, *n.*	3 Φύλλον, ου, n.
A Side	Pagĭna, æ, *f.*	5 Σελὶς ίδος, f.
A Margin	Margo, ĭnis, *f.*	3 Κράσπεδον, ου, n.
A Cover of a Book	Tegumentum, i, *n.*	5 Σκέπασμα, ατος, n.
A Printer	Typogrăphus, i, *m.*	3 Τυπογράφος, ου, m.
A Printing-Press	Prelum, i, *n.*	5 Πιεστὴρ, ῆρος, m.
A Composer	Typothĕta, æ, *m.*	1 Τυποθέτης, ου, m.
A Title of a Book	Inscriptio, ōnis, *f.*	2 Ἐπιγραφὴ, ῆς, f.
A Book-binder	Bibliopēgus, i, *m.*	3 Βιβλιοπηγὸς, οῦ, m.
A Book-seller	Bibliopōla, æ, *m.*	1 Βιβλιοπώλης, ου, m.
Paper	Charta, æ, *f.*	1 Χάρτης ου, m. Πάπυρος
Blotting Paper	Charta bibula	
A Sheet of Paper	Scheda, æ, *f.*	2 Σχέδη, ης, f.
A Quire / *A Ream* of Paper	Scapus, i, *m.*	
Parchment	Pergamēna, æ, *f.*	2 Περγαμῆναι ῶν, f.
A Manual	Enchiridium, ii, *n.*	3 Ἐγχειρίδιον, ίου, n.

A Volume

A Volume	Volūmen, ĭnis, *n.*	5 Σύνταγμα, ατος, n.
A Work	Opus, ĕris, *n.*	5 Σύγγραμμα, τος, n.
A Quill or Pen	Calămus, i, *m.* seu Penna scriptoria	3 Κάλαμος, άμου, m.
A Penknife	Scalpellum, i, *n.*	2 Σμίλη, ης, f.
A Pencil	Penicillus, i, *m.*	5 Γραφὶς, ίδος, f.
Ink	Atramentum, i, *n.*	5 Μέλαν, ανος, n.
An Inkhorn	Atramentarium, ii, *n*	2 Κίστη μελανδόχος
A Blot	Litūra, æ, *f.*	3 Σπίλος, ου, m.
Writing	Scriptio, ōnis, *f.*	2 Γραφὴ, ῆς, f.
A Character	Character, ēris, *m.*	5 Χαρακτὴρ, ῆρος, m.
A Point	Punctum, i, *n.*	2 Στιγμὴ, ῆς, f.
A Period	Periŏdus, i, *f.*	3 Περίοδος, όδου, f.
A Letter	Litĕra, æ, *f.*	5 Γράμμα, τος, n.
A Syllable	Syllăba, æ, *f.*	2 Συλλαβὴ, ῆς, f.
A Word	Vocabŭlum, i, *n.*	5 Ῥῆμα, τος, n. Λέξις εως
A Sentence	Sententia, æ, *f.*	2 Γνώμη, ης, f.
A Language	Lingua, æ, *f.*	2 Γλῶσσα, ης, f.
An Accent	Accentus, ûs, *m.*	2 Προσῳδία, ας, f.
A Spirit	Spirĭtus, ûs, *m.*	5 Πνεῦμα, τος, n.
A Figure	Figūra, æ, *f.*	5 Σχῆμα, τος, n.
A Lesson	Lectio, ōnis, *f.*	2 c. Ἀνάγνωσις, εως, f.
Construing	Interpretatio, ōnis, *f.*	2 Ἑρμηνεία, ας, f.
Parsing	Examinatio, ōnis, *f.*	2 Ἐξέτασις, εως, f.
A Rule	Regŭla, æ, *f.*	5 Κανὼν, όνος, m.
An Exception	Exceptio, ōnis, *f.*	2 c. Ἐξαίρεσις, εως, f.
An Exercise	Exercitium, ii, *n.*	5 Ἄσκημα, ατος, n.
A Theme	Thema, ătis, *n.*	5 Θέμα, τος, n.
Prose	Prosa, æ, *f.*	3 Λόγος πεζὸς, ου, m.
Verse	Carmen, ĭnis, *n.*	3 Λόγος ἔμμετρος
An Orator	Orātor, ōris, *m.*	5 Ῥήτωρ, ορος, m.
A Speech or Oration	Oratio, ōnis, *f.*	3 Λόγος, ου, m.
A Preface	Exordium, ii, *n.*	3 Προοίμιον, ίου, n.
A Confirmation	Confirmatio, ōnis, *f.*	2 Κατασκευὴ, ῆς, f.
A Confutation	Confutatio, ōnis, *f.*	2 Ἀνασκευὴ, ῆς, f.

A Con-

A Conclusion	Conclusio, ōnis, *f.*	3 Ἐπίλογος, ου, m.
A Poet	Poëta, æ, *m.*	1 Ποιητὴς, οῦ, m.
Poetry	Poësis, is, *f.*	2 c. Ποίησις, εως, f.
A Poem	Poëma, ătis, *n.*	5 Ποίημα, τος, n.
A Proverb	Adagium, ii, *n.*	2 Παροιμία, ας, f.
A Chronicle	Annāles, ium, *m.*	3 Χρονικὰ, ῶν, n.
A Day-Book	Diarium, ii, *n.*	5 Ἐφημερὶς, ίδος, f.
A Calendar	Fasti, orum, *m.*	3 Ἡμερολογεῖον, ου, n.
A Fable or Tale	Fabŭla, æ, *f.*	3 Μῦθος, ου, m.
A Table-Book	Pugillāres, ium, *m.*	3 Παλίμψηστος, ου, m.
A Truant	Emansor, ōris, *m.*	1 Διαμελλητὴς, οῦ, m.
A Dunce	Hebes, etis, *com.*	
A Rod	Virga, æ, *f.*	3 Ῥάβδος, ου, f.
A Ferular or Palmer	Ferŭla, æ, *f.*	5 Νάρθηξ, ηκος, m.

XXVI.

Of the Church and Ecclesiastic matters	*De Ecclesia & rebus Ecclesiasticis.*	Περὶ Ἱεροῦ καὶ Ἐκκλησιαστικῶν.
Religion	Religio, ōnis, *f.*	2 Εὐσέβεια, ας, f.
Superstition	Superstitio, nis, *f.*	2 Δεισιδαιμονία, ας, f.
A Christian	Christiānus, i, *m.*	3 Χριστιανὸς, οῦ, m.
An Heretick	Hæretĭcus, i, *m.*	3 Αἱρετικὸς, οῦ, m.
A Schismatick	Schismatĭcus, i, *m.*	3 Σχισματικὸς, οῦ, m.
A Heathen or Pagan	Ethnĭcus, i, *m.*	3 Ἐθνικὸς, οῦ, m.
A Temple	Templum, i, *n.*	3 Ναὸς, οῦ, m.
A Church	Ecclesia, æ, *f.*	2 Ἐκκλησία, ας, f.
A Chappel	Sacellum, i, *n.*	3 Ναΐδιον, ίου, n.
A Chancel	Adytum, i, *n.*	3 Ἄδυτον, ύτου, n.
A Church-Porch	Vestibŭlum templi	3 Προπύλαιον, αίου, n.
A Church-Yard	Cœmeterium, ii, *n.*	3 Κοιμητήριον, ίου, n.
A Vestry	Sacrarium, ii, *n.*	3 Ἱεροφυλάκιον, ίου, n.
A Pew	Subsellium, ii, *n.*	3 Ἑδώλιον, ίου, n.
A Pulpit	Suggestum, i, *n.*	3 Ἀνάβαθρον, άθρου, n.

A Bell

A Bell	Campāna, æ, *f.*	5 Κώδων, ωνος, m.
An Altar	Ara, æ, *f.*	3 Βωμὸς, οῦ, m.
A Statue	Statua, æ, *f.*	5 Ἄγαλμα, ατος, n.
An Image	Imāgo, ĭnis, *f.*	5 Ἐ[illegible], ό[illegible], f.
A Congregation	Congregatio, ōnis, *f.*	2 Συν[illegible], f.
A Lay-man	Laicus, i, *m.*	3 Λ[illegible]
A Clergy-man	Clerĭcus, i, *m.*	3 Κλη[illegible]
A Priest	Sacerdos, ōtis, *com.*	3 c. Ἱερεὺς, έως, m.
A Parish	Parochia, æ, *f.*	2 Παροικία, ας, f.
A Chaplain	Sacellānus, i, *m.*	
A Deacon	Diacŏnus, i, *m.*	3 Διάκονος, όνου, m.
A Bishop	Episcŏpus, i, *m.*	3 Ἐπίσκοπος, όπου, m.
A Diocess	Diœcēsis, is, *f.*	2 c. Διοίκησις, εως, f.
An Elder	Presbyter, ĕri, *m.*	3 Πρεσβύτερος, έρου, m.
A Reader	Lector, ōris, *m.*	1 Ἀναγνώστης, ου, m.
A Preacher	Concionātor, ōris, *m.*	3 Δημηγόρος, ου, m.
A Sermon	Concio, ōnis, *f.*	2 Ὁμιλία, ας, f.
A Catechism	Catechismus, i, *m.*	3 Κατηχισμὸς, οῦ, m.
A Churchwarden or Sexton	Æditnus, i, *m.*	3 Νεωκόρος, ου, m.
The Scriptures	Scriptūræ, ārum, *f.*	2 Γραφαὶ, ῶν, f.
The Bible	Biblia, ōrum, *n.*	3 Βιβλία, ίων, n.
The Old / *The New* } *Testament*	Testamentum } Vetus / Novum	2 Διαθήκη } Παλαιὰ / Καινὴ
A Chapter	Caput, ĭtis, *n.*	3 Κεφάλαιον, ου, n.
A Verse	Versus, ûs, *m.*	5 Κόμμα, ατος, n.
A Prophet	Prophēta, æ, *m.*	1 Προφήτης, ου, m.
A Prophesie	Prophetīa, æ, *f.*	2 Προφητεία, ας, f.
An Apostle	Apostŏlus, i, *m.*	3 Ἀπόστολος, όλου, m.
An Euangelist	Euangelista, æ, *m.*	1 Εὐαγγελιστὴς, οῦ, m.
The Gospel	Euangelium, ii, *n.*	3 Εὐαγγέλιον, ίου, n.
A Saint	Sanctus, i, *m.*	3 Ἅγιος, ίου, m.
A Martyr	Martyr, yris, *com.*	5 Μάρτυς, υρος, com.
A Monk	Monăchus, i, *m.*	3 Μόναχος, άχου, m.
A Nun	Monialis, is, *f.*	2 Μονάστρια, ιας, f.
An Abbat	Abbas, ătis, *m.*	1 Κοινοβιάρχης, ου, m.

An Hermit	Eremīta, æ, *m.*	1 Ἐρημίτης, ου, m.
A Dean	Decānus, i, *m.*	
The Pope	Papa, æ, *m.*	1 Πάππας, ου, m.
Fasting	Jejunium, ii, *n.*	2 Νηστεία, ας, f.
Prayers	Preces, um, *f.*	2 Προσευχή, ῆς, f.
A Liturgy	Liturgia, æ, *f.*	2 Λειτουργία, ας, f.
Confeſſion	Confeſſio, ōnis, *f.*	2 c. Ἐξομολόγησις εως, f.
Petition	Petitio, ōnis, *f.*	2 c. Αἴτησις, εως, f.
Deprecation	Deprecatio, ōnis, *f.*	2 c. Δέησις, εως, f.
Thanksgiving	Gratiarum actio, ōnis, *f*	2 Εὐχαριστία, ας, f.
Bleſſing	Benedictio, ōnis, *f.*	2 Εὐλογία, ας, f.
Singing	Cantio, ōnis, *f.*	2 c. Ἆσις, εως, f.
A Pſalm	Pſalmus, i, *m.*	3 Ψαλμός, οῦ, m.
The Pſalter	Pſalterium, ii, *n.*	3 Ψαλτήριον, ίου, n.
An Hymn	Hymnus, i, *m.*	3 Ὕμνος, ου, m.
An Anthem	Antiphōna, æ, *f.*	2 Ἀντιφώνη, ης, f.
A Song	Cantilēna, æ, *f.*	2 Ὠδή, ῆς, f.
A Quire	Chorus, i, *m.*	3 Χορός, οῦ, m.
A Singing-man	Cantor, ōris, *m.*	1 Ψάλτης, ου, m.
An Organ	Orgănum, i, *n.*	3 Ὄργανον, άνου, n.
A Chanter	Præcentor, ōris, *m.*	3 Χορηγός, οῦ, m.
A Sacrament	Sacramentum, i, *n.*	3 Μυστήριον, ίου, n.
Baptiſm	Baptiſmus, i, *m.*	3 Βαπτισμός, οῦ, m.
A Font	Baptiſterium, ii, *n.*	3 Βαπτιστήριον, ίου, n.
A Godfather	Suſceptor, ōris, *m.*	
A Godmother	Suſceptrix, īcis, *f.*	
A Goſſip { *He*	Compăter, tris, *m.*	5 Συμπάτηρ, τρος, m.
A Goſſip { *She*	Commāter, tris, *f.*	5 Συμμήτηρ, τρος, f.
Circumciſion	Circumciſio, ōnis, *f.*	2 Περιτομή, ῆς, f.
A Sacrifice	Sacrificium, ii, *n.*	2 Θυσία, ας, f.
The Paſſover	Paſcha, ătis, *n.*	5 Πάσχα, τος, n.
The Euchariſt or Sacrament of the Lords Supper	Euchariſtia, æ, *f.*	2 Εὐχαριστία, ας, f.
Conſecration	Conſecratio, ōnis, *f.*	2 c. Καθιέρωσις, εως, f.
Communion	Communio, ōnis, *f.*	2 Κοινωνία, ας, f.

Excom-

Excommunication	Excommunicatio, ōnis	2 c. Ἀποκήρυξις, εως, f.
Absolution	Absolutio, ōnis, *f.*	2 c. Ἀπόλυσις, εως, f.
A Grave	Sepulchrum, i, *n.*	3 Τάφος, ου, m.
A Bier	Ferĕtrum, i, *n.*	3 Φέρετρον, έτρου, n.
A Coffin	Capŭlum, i, *n.*	2 Θήκη, ης, f.
A Herse	Cenotaphium, ii, *n.*	3 Κενοτάφιον, ίου, n.
A Monument	Monumentum, i, *n.*	3 Μνημεῖον, ου, n.
An Epitaph	Epitaphium, ii, *n.*	3 Ἐπιτάφιον, ίου, n.
A Funeral	Funus, ĕris, *n.*	2 Ἐκφορὰ, ᾶς, f.
Funeral-Rites	Exequiæ, arum, *f.*	5 Ἐναγίσματα, ων, n.
A Scutcheon	Tholus, i, *m.*	3 Θόλος, ου, m.
Burial	Sepultūra, æ, *f.*	3 Ἐνταφιασμὸς, οῦ, m.

XXVII.

Of Husbandry and Country-affairs.	*De Agricultura & rebus Rusticis.*	Περὶ Γεωργικῶν.
THe Country	RUs, rūris, *n.*	3 Ἀγρὸς, οῦ, m.
A Country-man	Rustĭcus, i, *m.*	3 Ἄγροικος, ου, m
A Husbandman	Agricŏla, æ, *com.*	3 Γεωργὸς, οῦ, m.
Ground	Fundus, i, *m.* }	3 Χωρίον, ίου, n. Κτῆμα, τος
A Farm	Prædium, ii, *n.* }	
A Field	Ager, gri, *m.*	3 Ἀγρὸς, οῦ, m.
A Landlord	Dominus fundi	5 Κτήτωρ, ορος, m. ἀγροῦ ἢ οἰκίας
A Tenant	Manceps, ĭpis, *c.* Colō-(nus, i, *m.*	Ὁ μισθωσάμενος οἶκον ἢ (ἀγρὸς
A Steward or Bailiff	Villĭcus, i, *m.*	3 Οἰκονόμος, ου, m.
A Labourer or Workman	Operarius, ii, *m.*	1 Ἐργάτης, ου, m.
Arable Land	Arvum, i, *n.*	2 Ἄρουρα, ας, f.
A Turf	Cespes, ĭtis, *m.* }	3 Βῶλος, ου, m.
A Clod	Gleba, æ, *f.* }	
Dung	Fimus, i, *m.*	3 Κόπρος, ου, m.
A Plowman	Arātor, ōris, *m.*	5 Ἀροτὴρ, ῆρος, m.
A Plow	Arātrum, i, *n.*	3 Ἄροτρον, ότρου, n.
The Plow-handle	Stiva, æ, *f.*	2 Ἐχέτλη, ης, f.

The

The Plow-ſhare	Vomis, ĕris, *m.*	2 c. Ὕνις, εως, f.
The Coulter	Culter, tri, *m.*	2 Μάχαιρα, ας f.
An Harrow	Occa, æ, *f.*	5 Βωλοκόπημα, ατος, n.
A Yoke	Jugum, i, *n.*	3 Ζυγὸς, ȣ̃, m.
A Goad	Stimŭlus, i, *m.*	3 Κέντρον, ȣ, n.
A Furrow	Sulcus, i, *m.*	5 Ἄυλαξ, ακος, f.
A Ridge	Lira, æ, *f.*	
A Sower	Seminātor, ōris, *m.*	5 Σπείρων, οντος, m.
Seed	Semen, ĭnis, *n.*	3 Σπόρος, ȣ, m.
Standing-corn	Seges, ĕtis, *f.*	3 Λήϊον ȣ, n.
An Ear of Corn	Spica, æ, *f.*	5 Στάχυς υος, m.
The Blade	Culmus, i, *m.*	3 Κάλαμος, άμȣ, m.
Harveſt	Meſſis, is, *f.*	1 c. Θέρος, εος ȣς, n.
A Sheaf of Corn	Faſcis ſpicarum	2 Δέσμη, ης, f.
A Shock of Corn	Meta, æ, *f.*	3 Σωρὸς, ȣ̃, m.
A Gleaning	Spicilegium, ii, *m.*	2 Σταχυολογία, ας, f.
A Reaper	Meſſor, ōris, *m.*	1 Θεριστὴς, ȣ̃, m.
A Sickle	Falx, cis, *f.*	3 Δρέπανον, άνȣ, n.
Stubble	Stipŭla, æ, *f.*	2 Καλάμη, ης, f.
A Wain	Vehicŭlum, i, *n.*	5 Ὄχημα, ατος, n.
A Cart	Plauſtrum, i, *n.*	2 Ἄμαξα, ης, f.
A Wheel	Rota, æ, *f.*	3 Τροχὸς ȣ̃, m.
A Spoke	Radius, ii, *m.*	2 Κνημία, ας, f.
An Axle-tree	Axis, is, *m.*	5 Ἄξων, ονος, m.
A Carter or Wagoner	Aurīga, æ, *com.*	3 Ἡνίοχος, ȣ, m.
A Whip	Scutĭca, æ, *f.*	5 Σκυταλὶς, ίδος, f.
The Reins	Habēna, æ, *f.*	2 Ἡνία, ας, f.
A Barn	Horreum, i, *n.*	2 Ἀποθήκη, ης, f.
A Threſhing-floor	Area, æ, *f.*	5 Ἅλων, ωνος, f.
A Threſher	Tritor, ōris, *m.*	5 Ἀλοητὴρ, ῆρος, m.
A Flail	Tribŭla, æ, *f.*	5 Σιτότριψ, ϐος, m.
A Fan	Vannus, i, *m.*	3 Λικμὸς, ȣ̃, m.
A Sieve	Cribrum, i, *n.*	3 Κόσκινον ίνȣ, n.
Straw	Stramen, ĭnis, *n.*	2 Κάρφη, ης, f.
A Grain	Granum, i, *n.*	3 Κόκκος ȣ, m.

A Granary	Granarium, ii, *n.*	5 Σιτοβολὼν, ῶνος, m.
A Sack	Saccus, i, *m.*	3 Σάκκος, ου, m.
Bread-corn	Frumentum, i, *n.*	3 Σῖτος, ου, m.
Mault	Braſium, ii, *n.* Byne, es, *f.*	2 Βυύη, ης, f.
A Paſture	Paſcuum, i, *n.*	2 Νομὴ, ῆς, f.
A Meadow	Pratum, i, *n.*	5 Λειμὼν, ῶνος, m.
Hay	Fœnum, i, *n.*	3 Χόρτος, ου, m.
Hay-Harveſt	Fœniſecium, ii, *n.*	2 Χορτοτομία, ας, f.
A Mower of Hay	Fœnisĕca, æ, *m.*	3 Χορτοτόμος, ου, m.
A Hay-loft	Fœnīle, is, *n.*	3 Χορτοδοχεῖον, ου, n.
A Rake	Raſtrum, i, *n.*	2 Ἀγρείφνη, ης, f.
A Cart or Wain-Load	Vehes, is, *f.*	5 Ὄχημα, ατος, n.
A Bundle or Faggot	Faſcis, is, *m.*	2 Δέσμη, ης, f.
A Fork	Furca, æ, *f.*	2 Δίκελλα, ης, f.
A Fork with three Tines	Tridens, tis, *m.*	2 Τρίαινα, ης, f.
A Garden	Hortus, i, *m.*	3 Κῆπος, ου, m.
A Gardener	Hortulānus, i, *m.*	3 Κηπουρὸς, οῦ, m.
A Garden of Pleaſure	Paradīſus, i, *f.*	3 Παράδεισος, ου, f.
An Orchard	Pomarium, ii, *n.*	5 Μηλεὼν, ῶνος, m.
A Bank	Agger, ĕris, *m.*	5 Χῶμα, ατος, n.
A Wall	Murus, i, *m.*	1 c. Τεῖχος, εος ους, n.
A Hedge	Sepes, is, *f.*	3 Φραγμὸς, οῦ, m.
A Ditch	Foſſa, æ, *f.*	3 Τάφρος, ου, m.
A Digger	Foſſor, ōris, *m.*	3 c. Σκαπανεὺς, έως, m.
An Arbour	Topiarium, ii, *n.*	
A Bed in a Garden	Pulvillꝰ, i, *m.* Areŏla, æ, *f.*	2 Πρασιὰ, ᾶς, f.
A Spade	Ligo, ōnis, *m.*	2 Μάκελλα, ας, f.
A Shovel	Pala, æ, *f.*	2 Σκαπάνη, ης, f.
A How	Sarcŭlum, i, *n.*	3 Σκαλιστήριον, ίου, n.
A Pick-ax	Rutrum, i, *n.*	3 Σκαφεῖον, ου, n.
A Mattock	Bipalium, ii, *n.*	3 Μάῤῥον, ου, n.
A Pruning-hook	Falx putatoria	3 Κλαδευτήριον, ίου, n.
A Prop	Fulcrum, i, *n.*	5 Ἕρμα, ατος, n.
A Rowler	Cylindrus, i, *m.*	3 Κύλινδρος, ου, m.
A Wheel-barrow	Vehicŭlum truſatile	

A Hand-barrow	Vehicŭlum manuale	
A Wood	Sylva, æ, *f.*	2 Ὕλη, ης, f.
A Grove	Lucus, i, *m.*	1 c. Ἄλσος, εος, ους, n.
A Nursery	Seminarium, ii, *n.*	3 Φυτευτήριον, ίου, n.
A Park	Vivarium ferarum	3 Θηριοτροφεῖον, ου, n
An Heath	Ericētum, i, *n.*	5 Ἐρεικεών, ῶνος, m.

XXVIII.

Of Warfare.	*De Re Militari.*	Περὶ Στρατιωτικῶν.
WAR	BEllum, i, *n.*	3 ΠΟ'λεμος, ου, m.
Peace	Pax, pacis, *f.*	2 Εἰρήνη, ης, f.
A Truce	Induciæ, arum, *f.*	2 Ἐκεχειρία, ας, f.
Warfare	Militia, æ, *f.*	2 Στρατεία, ας, f.
An Enemy	Hostis, is, *m.*	3 Πολέμιος, ίου, m.
A Friend	Socius, ii, *m.*	3 Σύμμαχος, άχου, m.
Pay	Stipendium, ii, *n.*	3 Μισθὸς, οῦ, m.
Forces	Copiæ, arum, *f.*	2 c. Δύναμις, εως, f.
An Army	Exercitus, ûs, *m.*	3 Στρατὸς, οῦ, m.
A Regiment	Legio, ōnis, *f.*	5 Λεγεών, ῶνος, m.
A Troop of Horse	Turma, æ, *f.*	2 Ἴλη, ης, f.
A Company of Foot	Cohors, tis, *f.* Centuria	3 Λόχος, ου, m.
A Band or Squadron of Soldiers	Manipŭlus, i, *m.*	
A Brigade	Agmen, ĭnis, *n.*	5 Τάγμα, ατος, n.
A General	Imperātor, ōris, *m.*	3 Στρατηγὸς, οῦ, m.
A Colonel	Tribūnus, i, *m.*	3 Χιλίαρχος, ου, m.
A Captain	Dux, dŭcis, *m.*	3 Λοχαγὸς, οῦ, m.
A Captain of an hundred	Centurio, ōnis, *m.*	3 Ἑκατόνταρχος, ου, m.
A Lieutenant	Locum tenens	
A Cornet or Ensign	Vexillĭfer, ĕri, *m.* Signĭfer	3 Σημειοφόρος, ου, m.
A Banner or Colours	Vexillum, i, *n.* Signū, i, *n.*	3 Σημεῖον, ου, n.
A Trumpeter	Tubĭcen, ĭnis, *m.*	1 Σαλπιγκτὴς, οῦ, m.
A Drummer	Tympanista, æ, *m.*	1 Τυμπανιστὴς, οῦ, m.
A Serjeant	Decurio, ōnis, *m.*	1 Δεκαδάρχης, ου, m.

A Quarter-maſter	Menſor, ōris, *m.*	3 Ἐπίσταθμος, ου, m.
The Round	Circĭtores, um, *m.*	
A Soldier	Miles, ĭtis, *m.*	1 Στρατιώτης, ου, m.
A common Soldier	Miles gregarius	1 Ἐγκύκλιος, ίου, m.
A Horſeman or Trooper	Eques, ĭtis, *m.*	3 c. Ἱππεὺς, έως, m.
A Footman	Pedes, ĭtis, *m.*	3 Πεζὸς, οῦ m.
The Cavalry	Equitātus, ûs, *m.*	2 Ἱππικὴ, ῆς, f.
The Infantry	Peditātus, ûs, *m.*	2 Πεζικὴ, ῆς, f. δύναμις
An Archer	Sagittarius, ii, *m.*	1 Τοξότης, ου, m.
A Pikeman	Haſtātus, i, *m.*	3 Δορυφόρος, ου, m.
A Muſketeer	Sclopetarius, ii, *m.*	
A Scout or Spy	Explorātor, ōris, *m.*	3 Κατάσκοπος, ου, m.
A Sentinel	Stationarius miles	5 Προφύλαξ, ακος, m.
The Guard or Watch	Excubiæ, arum, *f.*	2 Φυλακαὶ, ῶν, f.
A Battel	Prælium, ii, *n.*	2 Συμβολὴ, ῆς, f.
A Fight	Pugna, æ, *f.*	2 Μάχη, ης, f.
Challenging	Provocatio, ōnis, *f.*	2 c. Πρόκλησις, εως, f.
A Duel	Monomachia, æ, *f.*	2 Μονομαχία, ας, f.
An Aſſault	Impreſſio, ōnis, *f.*	2 c. Ἐπιχείρησις, εως, f.
An Expedition	Expeditio, ōnis, *f.*	2 Στρατεία, ας, f.
A Skirmiſh	Velitatio, ōnis, *f.*	3 Ἀκροβολισμὸς, οῦ, m.
A Retreat	Receptus, ûs, *m.*	2 c. Ἀναχώρησις, εως, f.
A Flight	Fuga, æ, *f.*	2 Φυγὴ, ῆς, f.
An Ambuſh	Inſidiæ, arum, *f.*	3 Λόχος, ου, m. Ἐνέδρα, ας
A Victory	Victoria, æ, *f.*	2 Νίκη, ης, f.
A Slaughter	Strages, is, *f.*	2 Κοπὴ, ῆς, f.
A Priſoner	Captīvus, i, *m.*	3 Αἰχμάλωτος, ου, m.
Plundering	Direptio, ōnis, *f.*	3 Ἁρπαγμὸς, οῦ, m.
Spoil	Spolia, orum, *n.*	3 Σκῦλον, ου, n. Λάφυρα, ων, n.
Prey or Booty	Præda, æ, *f.*	2 Λεία, ας, f.
A Siege	Obſidio, ōnis, *f.*	2 c. Πολιόρκησις, εως, f.
A Mine	Cunicŭlus, i, *m.*	3 Ὑπόνομος, ου, m.
An Hoſtage	Obſes, ĭdis, *com.*	3 Ὅμηρος, ήρου, m.
The Camp	Caſtra, orum, *n.*	3 Στρατόπεδον, ου, n.
A Garriſon	Præſidium, ii *n.*	2 Φρουρὰ, ᾶς, f.

A Bul-

A Bulwark	Agger, ĕris, *m.*	5 Χῶμα, ατος, n.
A Fort or Outwork	Propugnacŭlum, i, *n.*	5 Προτείχισμα, ατος, n.
A Tent	Tentorium, ii, *n.*	2 Σκηνή, ης, f.
Forage	Pabŭlum, i, *n.*	2 Νομὴ, ῆς, f.
A Sword	Gladius, ii, *m.*	2 Μάχαιρα, ας, f.
A Dagger	Pugio, ōnis, *f.*	3 Ἐγχειρίδιον, ίου, n.
A Scabbard	Vagīna, æ, *f.*	3 Κολεὸς, οῦ, m.
A Belt	Baltheus, i, *m.*	5 Ζωστὴρ, ῆρος, m.
A Dart	Telum, i, *n.*	1 c. Βέλος, εος ους, n.
A Pike	Hasta, æ, *f.*	5 Δόρυ, ατος & δουρός, n.
An Halberd	Bipennis, is, *f.*	5 Ἀμφιπέλεκυς, εως, f.
A Bow	Arcus, ûs, *m.*	3 Τόξον, ου, n.
A Bow-string	Nervus, i, *m.*	3 Σύνδεσμος, ου, m.
An Arrow	Sagitta, æ, *f.*	3 Ἰὸς, οῦ, m.
A Quiver	Pharĕtra, æ, *f.*	2 Φάρετρα, ας, f.
A Sling	Funda, æ, *f.*	2 Σφενδόνη, ης, f.
A Club	Clava, æ, *f.*	3 Ῥόπαλον, άλου, n.
A Piece of Ordnance	Tormentum majus	
A Gun	Bombarda, æ, *f.*	
A Musket	Sclopētum, i, *n.*	
A Bullet	Glans, dis, plumbea	
Match	Fomes igniarius	
Gun-Powder	Pulvis tormentarius aut pyrius	
An Helmet or Head-piece	Galea, æ, *f.*	5 Κόρυς, υθος, f.
A Brest-plate	Thorax, ācis, *m.*	5 Θώραξ, ακος, m.
A Buckler	Clypeus, i, *m.*	5 Ἀσπὶς, ίδος, f.
A Shield	Scutum, i, *n.*	3 Θυρεὸς, οῦ, m.
Disbanding or Casheering	Missio, ōnis, *f.*	
A Muster	Armilustrium, ii, *n.*	

XXIX. Of

XXIX.

Of Shipping and Navigation.	*De Re Navali.*	Περὶ Ναυτικῶν.
A Navy	Classis, is, *f.*	3 Στόλος, ου, m.
A Ship	Navis, is, *f.*	5 Ναῦς, ναός, f.
A Merchant-man	Navis oneraria, *f.*	3 Πλοῖον φορτικόν
A Man of War	Navis bellica	
A Galley	Navis longa, Triremis, is, *f.*	1 c. Τριήρης, εος, ους, f. sub. ναῦς
A Skiff or Cock-boat	Scapha, æ, *f.*	2 Σκάφη, ης, f.
A Fisher-boat	Cymba, æ, *f.*	2 Κύμβη, ης, f.
A Boat	Navicŭla, æ, *f.*	3 Πλοιάριον, ίου, n.
A Ferry-boat	Ponto, ōnis, *m.*	3 Πορθμεῖον, ου, n.
A Ferry-man	Portĭtor, ōris, *m.*	3 c. Πορθμεύς, έως, m.
The Keel of a Ship	Carīna, æ, *f.*	5 Τρόπις, ιδος, f.
The Stem or Prow	Prora, æ, *f.*	2 Πρῶρα, ας, f.
The Stern or Poop	Puppis, is, *f.*	2 Πρύμνη, ης, f.
The Hull or Hulk	Alveus, i, *m.*	
The Helm or Rudder	Clavus, i, *m.* Gubernacŭlũ	5 Οἴαξ, ακος, m.
The Decks	Fori, ōrũ, *m.* Tabulatũ (navis	5 Κατάστρωμα, ατος, n.
A Rower	Remex, ĭgis, *c.*	1 Ἐρέτης, ου, m.
An Oar	Remus, i, *m.*	2 Κώπη, ης, f.
A Sail	Velum, i, *n.*	3 Ἱστίον, ου, n.
The Sail-yard	Antenna, æ, *f.*	5 c. Κέρας, ατος, αος, ως, n.
The Mast	Malus, i, *m.*	3 Ἱστὸς, οῦ, m.
The Main-Sail	Artĕmon, ŏnis, *m.*	5 Ἀρτέμων, ονος, m.
The Fore-Sail	Dolon, ōnis, *m.*	5 Δόλων, ωνος, m.
The Mizzen-Sail	Epidrŏmus, i, *m.*	3 Ἐπίδρομος, ου, c.
A Rope	Funis, is, *m.*	3 Σπάρτον, ου, n. Σχοινίον
A Cable	Rudens, tis, *dub.*	3 Κάλος, ου, m.
An Anchor	Anchŏra, æ, *f.*	2 Ἄγκυρα, ύρας, f.
Ballast	Saburra, æ, *f.*	5 Ἕρμα, ατος, n.

The

The Master of a Ship	Nauclērus, i, *m.*	3 Ναύκληρος, ήρου, m.
The Pilot	Gubernātor, ōris, *m.*	1 Κυβερνήτης, ου, m.
A Mariner	Nauta, æ, *m.*	1 Ναύτης, ου, m.
A common Seaman	Socius navalis	
A Passenger	Vector, ōris, *m.*	1 Ἐπιβάτης, ου, m.
Fraight	Naulum, i, *n.*	3 Ναῦλον, ου, n.
Shipwrack	Naufragium, ii, *n.*	3 Ναυάγιον, ίου, n.
A Shipwright	Naupēgus, i, *m.*	3 Ναυπηγὸς, οῦ, m.
A Sea-fight	Naumachia, æ, *f.*	2 Ναυμαχία, ας, f.
An Admiral	Archithalassus, i, *m.*	3 Ἀρχιθάλασσος, ου, m.
The Cargo or Lading of a Ship	Onus, ĕris, *n.*	3 Γόμος, ου, m.
A Merchant	Mercātor, ōris, *m.*	3 Ἔμπορος, όρου, m.
The Sink	Sentīna, æ, *f.*	
The Compass	Pyxidŭla nautĭca, *f.*	5 Πύξις, ιδος, f.
The Sounding Line	Bolis, ĭdis, *f.*	5 Βολὶς, ίδος, f.
The Flag	Aplustre, is, *n.*	

XXX.

Of Arts Liberal and Mechanic.	*De Artibus Liberalibus & Mechanicis.*	Περὶ Τεχνῶν ἐλευθερίων καὶ βαναυσῶν.
Philosophy	Philosophia, æ, *f.*	2 Φιλοσοφία, ας, f.
A Philosopher	Philosophus, i, *m.*	3 Φιλόσοφος, ου, m.
Grammar	Grammatĭca, æ, *f.*	2 Γραμματικὴ, ῆς, f.
Logick	Logĭca, æ, *f.*	2 Διαλεκτικὴ, ῆς, f.
A Logician	Logĭcus, i, *m.*	3 Διαλεκτικὸς, οῦ, m.
Rhetorick	Rhetorĭca, æ, *f.*	2 Ῥητορικὴ, ῆς, f.
Eloquence	Eloquentia, æ, *f.*	5 Λογιότης, ητος, f.
Musick	Musĭca, æ, *f.*	2 Μουσικὴ, ῆς, f.
Harmony	Harmonia, æ, *f.*	2 Ἁρμονία, ας, f.
A Note or Tune	Tonus, i, *m.*	3 Τόνος, ου, m.
A Fiddle	Fides, dis, *f.*	5 Χέλυς, υος, f.

A Fiddle-

A Fiddle-string	Chorda, æ, *f.*	2 Χορδὴ, ῆς, f.
A Quill	Plectrum, i, *n.*	3 Πλῆκτρον, ου, n.
A Lute	Cithăra, æ, *f.*	2 Κιθάρα, ας, f.
A Viol	Pandūra, æ, *f.*	2 Πανδοῦρα, ας, f.
* *A Harp*	Lyra, æ, *f.*	2 Λύρα, ας, f.
A Pipe	Tibia, æ, *f.*	3 Αὐλὸς, οῦ, m.
A Piper	Tibīcen, ĭnis, *m.*	1 Αὐλητὴς, οῦ, m.
A Whistle	Fistŭla, æ, *f.*	5 Σύριγξ, γγος, f.
A Taber or Drum	Tympănum, i, *n.*	3 Τύμπανον, άνου, n.
Arithmetick	Arithmetĭca, æ, *f.*	2 Ἀριθμητικὴ, ῆς, f.
A Number	Numĕrus, i, *m.*	3 Ἀριθμὸς, οῦ, m.
Astronomy	Astronomia, æ, *f.*	2 Ἀστρονομία, ας, f.
An Astronomer	Astronŏmus, i, *m.*	3 Ἀστρονόμος, ου, m.
Astrology	Astrologia, æ, *f.*	2 Ἀστρολογία, ας, f.
An Astrologer	Astrolŏgus, i, *m.*	3 Ἀστρολόγος, ου, m.
A Lawyer	Jurisconsultus, i, *m.*	3 Νομικὸς, οῦ, m.
A Sophister	Sophista, æ, *m.*	1 Σοφιστὴς, οῦ, m.
A Stage-player	Histrio, ōnis, *m.*	1 Ὑποκριτὴς, οῦ, m.
A Stage	Theātrum, i, *n.*	3 Θέατρον, άτρου, n.
A Vizard	Larva, æ, *f.*	3 Προσωπεῖον, ου, n.
A Comedy	Comœdia, æ, *f.*	2 Κωμῳδία, ας, f.
A Comedian	Comĭcus, i, *m.*	3 Κωμῳδοποιὸς, οῦ, m.
A Tragedy	Tragœdia, æ, *f.*	2 Τραγῳδία, ας, f.
A Tragedian	Tragĭcus, i, *m.*	3 Τραγῳδοποιὸς, οῦ, m.
A History	Historia, æ, *f.*	2 Ἱστορία, ας, f.
A Historian	Historĭcus, ci, *m.*	3 Ἱστορικὸς, οῦ, m.
An Artificer or Tradesman	Artĭfex, ĭcis, *m.*	1 Τεχνίτης, ου, m.
A Work-man or Handicrafts-man	Opĭfex, ĭcis, *m.*	1 Χειροτέχνης, ου, m.
A Farrier	Veterinarius, ii, *m.*	3 Κτηνίατρος, άτρου, m.
A Hatter or Capper	Coactiliarius, ii, *m.*	3 Πιλοποιὸς, οῦ, m.
A Goldsmith	Aurĭfex, ĭcis, *m.*	3 Χρυσουργὸς, οῦ, m.

* H. *Stephen* makes *Psalterium*, which we English a Psaltery, to be the same Instrument which we in English call a Harp.

A Carpenter

* *A Carpenter*	Faber lignarius	5 Τέκτων, ονος, m.
* *A Blackſmith*	Faber ferrarius	3 c. Χαλκεὺς έως, m.
A Hammer	Malleus, i, *m.*	2 Σφῦρα, ας f.
An Anvil or Stithy	Incus, ūdis, *f.*	5 Ἄκμων, ονος, f.
A Smiths Vice or A Screw	Cochlea, æ, *f.*	1 Κοχλίας, ου, m.
A File	Lima, æ, *f.*	2 Ῥίνη, ης, f.

** After the ſame manner ſeveral other Trades and Manufactures are denominated from the ſubject or matter about which they are converſant; as,

From:

Cuprum ſignifying Copper,	a Copper-ſmith may be called *Faber Cuprarius.*
Æs,	a Braſier *Ærarius.*
Stannum,	a Pewterer *Stannarius.*
Plumbum,	a Plummer *Plumbarius.*
Arma,	an Armourer *Armamentarius.*
Vitrum,	a Glaſier *Vitrarius.*
Culter,	a Cutler *Cultrarius.*
Gladius,	a Sword-ſmith *Gladiarius.*
Clavis,	a Lock-ſmith *Claviarius.*
Materia ſignifying Timber,	a Woodmonger *Materiarius.*
Reſtis,	a Rope-maker *Reſtiarius* and *Reſtio.*
Candela,	a Chandler *Candelarius.*
Vinum,	a Vintner *Vinarius.*
Popina,	a Victualler *Popinarius.*
Avis,	a Poulterer *Aviarius.*
Pomum,	a Coſtard-monger *Pomarius.*
Oleum,	an Oil-ſeller *Olearius.*
Farina,	a Meal-man *Farinarius.*
Sericum,	a Silk-man *Sericarius.*
Aromata,	a Grocer *Aromatarius.*
Zona,	a Girdler *Zonarius.*
Chirotheca,	a Glover *Chirothecarius.*
Caliga,	a Hoſier *Caligarius.*
Sella,	a Sadler *Sellarius.*
Carbo,	a Collier *Carbonarius.*
Vimen,	a Basket-maker *Viminarius.*
Cribrum,	a Sieve-maker *Cribrarius.*
Pecten,	a Comb-maker *Pectinarius.*

And many ſuch like.

A Cooper	Victor, ōris, *m.*	3 Οἰσυουργὸς, οῦ, m.
An Ax or Hatchet	Secūris, is, *f.*	5 Πέλεκυς, εως, m.
A Plain	Dolābra, æ, *f.* Planŭla, æ.	3 Ξύστρον, ου, n.
A Saw	Serra, æ, *f.*	2 c. Πρίσις, εως, f.
A Wedge	Cuneus, i, *m.*	5 Σφὴν, ηνός, m.
A Pair of Compasses	Circĭnus, i, *m.*	1 Διαβήτης, ου, m.
A Square	Norma, æ, *f.*	5 Γνώμων, ονος, c.
A Ruler	Regŭla, æ, *f.*	5 Κανὼν, όνος, m.
A Wimble or Auger	Terĕbra, æ, *f.*	3 Τρύπανον, άνου, n.
A Leaver	Vectis, is, *m.*	3 Μοχλὸς, οῦ, m.
Glue	Gluten, ĭnis, *n.*	2 Κόλλα, ης, f.
A Stone-cutter	Lapicīda, æ, *m.*	3 Λατόμος, ου, m.
A Chisel or Graving-tool	Celtis, is, *f.*	3 Γλυφεῖον, ου, n.
A Trowel	Trulla, æ, *f.*	
A Painter	Pictor, ōris, *m.*	3 c. Γραφεὺς, έως, m.
A Pencil	Penicillus, i, *m.*	3 Γραφεῖον, ου, n.
A Potter	Figŭlus, i, *m.*	3 c. Κεραμεὺς, έως, m.
A Butcher	Lanius, ii, *m.*	3 Κρεουργὸς, οῦ, m.
A Cook	Coquus, i, *m.*	3 Μάγειρος, είρου, m.
A Baker	Pistor, ōris, *m.*	3 Ἀρτοκόπος, ου, m.
A Miller	Molĭtor, ōris, *m.*	3 Μυλωθρὸς, οῦ, m.
A Fuller	Fullo, ōnis, *m.*	3 c. Γναφεὺς, έως, m.
A Dier	Tinctor, ōris, *m.*	3 c. Βαφεὺς, έως, m.
A Weaver	Textor, ōris, *m.*	1 Ὑφάντης, ου, m.
A Loom	Textrīnum, i, *n.*	3 Ἱστὸς, οῦ, m.
A Web	Tela, æ, *f.*	5 Ὕφασμα, ατος, n.
The Warp	Stamen, ĭnis, *n.*	5 Στήμων, ονος, m.
The Woof	Subtegmen, ĭnis, *n.*	2 Κρόκη, ης, f.
A Yarn-windle	Alabrum, i, *n.*	
A Furrier	Pellio, ōnis, *m.*	
A Tanner	Coriarius, ii, *m.*	1 Βυρσοδέψης, ου, m.
A Shoo-maker	Sutor calcearius, *m.*	3 Ὑποδηματοῤῥάφος, ου.
A Cobler	Cerdo, ōnis, *m.*	3 Σκυτοτόμος, ου, m.
A Horse-courser	Hippoplānus, i, *m.*	3 Ἱπποπλάνος, ου, m.
A Post	Veredarius, ii, *m.*	3 Ἡμεροδρόμος, ου, m.

A Letter-

A Letter-Carrier	Tabellarius, ii, *m.*	3 Γραμματοφόρος, ου, m.
A Porter	Bajŭlus, i, *m.*	5 Φόρταξ, ακος, m.
A Barber	Tonſor, ōris, *m.*	3 c. Κουρεὺς, έως, m.
A Barbers Shop	Tonſtrīna, æ, *f.*	3 Κουρεῖον, ου, n.
A Laundreſs	Lotrix, īcis, *f.*	2 Πλύντρια, ίας, f.
A Mountebank	Medĭcus circŭforaneus	1 Περιοδευτὴς, οῦ, m.
A Jugler	Præſtigiātor, ōris, *m.*	5 Γόης, ητος, m.
A Dancer	Saltātor, ōris, *m.*	1 Χορευτὴς, οῦ, m.
A Wreſtler	Luctātor, ōris, *m.*	1 Παλαιστὴς, οῦ, m.
A Rope-Dancer	Funambŭlus, i, *m.*	1 Σχοινοβάτης, ου, m.

XXXI.

Of Time and its meaſures.	*De Tempore ejuſq; menſuris.*	Περὶ Χρόνου καὶ τῶν αὐτοῦ μέτρων.
Time	Tempus, ŏris, *n.*	3 Χρόνος, ου, m.
A Seaſon	Tempeſtas, ātis, *f*	2 Ὥρα, ας, f.
Opportunity	Opportunĭtas, ātis, *f.*	3 Καιρὸς, οῦ, m.
A Year	Annus, i. *m.*	3 Ἐνιαυτὸς, οῦ, m.
Two Years	Biennium, ii, *n.*	1 c. Διετὲς, έος, οῦς, n.

So onwards three years is *Triennium*, four years *Quadriennium*, five years *Quinquennium*, *&c.*

Four Years	Luſtrum, i, *n.*	5 Ὀλυμπιὰς, άδος, f.
Leap-Year	Biſſextīlis annus, *m.*	
The Spring	Ver, veris, *n.*	5 Ἔαρ, αρος, n.
The Summer	Æſtas, ātis, *f.*	1 c. Θέρος, εος, ους, n.
The Autumn	Autumnus, i, *m.*	2 Ὀπώρα, ας, f.
The Winter	Hyems, is, *f.*	5 Χειμὼν, ῶνος, m.
A Month	Menſis, is, *m.*	5 Μὴν, μηνός, m.
January	Januarius, ii, *m.*	5 Γαμηλιὼν, ῶνος, m.
February	Februarius, ii, *m.*	5 Ἐλαφηβολιὼν, ῶνος, m
March	Martius, ii, *m.*	5 Μουνυχιὼν, ῶνος, m.

April	Aprīlis, is, *m.*	5 Θαργηλιὼν, ῶνος, m.
May	Maius, ii, *m.*	5 Σκιῤῥοφοριὼν ῶνος, m.
June	Junius, ii, *m.*	5 Ἑκατομβαιὼν, ῶνος, m.
July	Julius, ii, *m.*	5 Μεταγειτνιὼν, ῶνος, m.
August	Auguſtus, i, *m.*	5 Βοηδρομιὼν, ῶνος, m.
September	September, bris, *m.*	5 Μαιμακτηριὼν ῶνος, m
October	October, bris, *m.*	5 Πυανεψιὼν, ῶνος, m.
November	November, bris, *m.*	5 Ἀνθεστηριὼν ῶνος, m.
December	December, bris, *m.*	5 Ποσειδεὼν, ῶνος, m.
A Week	Septimāna, æ, *f.*	5 Ἑβδομὰς, άδος, f.
A Day	Dies, ei, *dub.*	2 Ἡμέρα, ας, f.
Sunday	Dies Dominĭcus, i, vel Solis	Ἡμέρα Κυριακὴ ἢ πρώτη
Munday	Dies Lunæ	Ἡμέρα δευτέρα
Tueſday	Dies Martis	Ἡμέρα τρίτη
Wedneſday	Dies Mercurii	Ἡμέρα τετάρτη
Thurſday	Dies Jovis	Ἡμέρα πέμπτη
Friday	Dies Veneris	Ἡμέρα ἕκτη
Saterday	Dies Saturni aut Sabbat	Ἡμέρα ἑβδόμη ἢ σαββάτων
Break of day	Dīlucŭlum, i, *n.*	3 Ὄρθρος, ου, m.
The Morning	Aurōra, æ, *f.*	4 c. Ἠὼς, όος, οῦς, f.
Noon	Meridĭes, ei, *m.*	2 Μεσημβρία, ας, f.
The Afternoon	Temp⁹ pomeridiānum	2 Δείλη, ης, f.
The Evening	Veſper, ĕris, *m.*	2 Ἑσπέρα, ας, f.
The Twilight	Crepuſcŭlum, i, *n.*	2 Ἀμφιλύκη, ης, f.
The Sun-riſing	Exortus Solis	2 Ἀνατολὴ, ῆς, f.
The Sun-ſetting	Occaſus, ûs, *m.*	2 Δυσμὴ, ῆς, f.
A Night	Nox, noctis, *f.*	5 Νὺξ, κτὸς, f.
Midnight	Nox intempeſta	3 Μεσονύκτιον, ίου, n.
Cock-crowing	Gallicinium, ii, *n.*	2 Ἀλεκτρυοφωνία, ας, f.
An Hour	Hora, æ, *f.*	2 Ὥρα, ας, f.
Half an hour	Semihōra, æ, *f.*	3 Ἡμιώριον, ίου, n.
A Quarter of an hour	Quadrans, tis, *m.*	3 Τεταρτημόριον, ίου, n.
A Minute	Minūtum, i, *n.*	

A Mo-

A Moment	Momentum, i, *n.*	2 Στιγμὴ, ῆς, f. Ἄτομον
An Hour-glass	Clepsydra, æ, *f.*	2 Κλεψύδρα, ας, f.
A Clock	Horologium, ii, *n.*	3 Ὡρολόγιον, ίου, n.
A Sun-Dial	Horologium solarium aut sciothericum	3 Ὡρολόγιον σκιοθηρικόν
The Point of a Dial	Stylus, i, *m.*	5 Γνώμων, ονος, m.
A Working-day	Dies profestus, i, *m.*	3 Ἀνέορτος, ου, com.
An Holy-day	Dies festus, i, *m.*	2 Ἑορτὴ, ῆς, f.
An half Holy-day	Dies intercisus	
Christmas	Natālis Christi	3 Γενέθλια, ίων, n.
Easter	Pascha, ătis, *n.*	5 Πάσχα, n. indec.
Whitsuntide	Pentecoste, es, *f.*	2 Πεντεκοστὴ, ῆς, f.
Infancy	Infantia, æ, *f.*	5 Νηπιότης, ότητος, f.
An Infant	Infans, tis, *com.*	3 Νήπιος, ίου, m.
Childhood	Pueritia, æ, *f.*	2 Παιδία, ας, f.
A Boy	Puer, ĕri, *m.*	5 Παῖς, παιδός, m.
A Girl or Wench	Puella, æ, *f.*	2 Παιδίσκη, ης, f.
A Youth	Adolescens, tis, *c.*	3 Ἔφηβος, ήβου, m.
A Young-man	Juvĕnis, is, *m.*	1 Νεανίας, ίου, m.
Youth	Juventus, ūtis, *f.*	2 Ἥβη, ης, f.
Manhood	Virilis ætas	Ἡλικία μέση ἢ ἀνδρῶν
Old-Age	Senectus, ūtis, *f.*	5 c. Γῆρας, ατος αος ως, n.
An Old-man	Senex, is, *m.*	5 Γέρων, οντος, m.
An Old-woman	Anus, ûs, *f.*	5 Γραῦς, αός, f.
An Age	Secŭlum, i, *n.*	5 Αἰὼν, ῶνος, m.
The Age of Man	Ætas, ātis, *f.*	2 Ἡλικία, ας, f.

XXXII.

Of Number.	*De Numero.*	Περὶ Ἀριθμοῦ.
Number	Numĕrus, i, *m.*	3 Ἀριθμὸς, οῦ, m.
Even	Par, paris, *m.*	3 Ἄρτιος, ίου, m
Odd	Impar, ris, *m.*	3 Περιττὸς, οῦ, m.

One,

One, 1, I.	Unus, a, um	Εἷς, μία, ἕν
The First	Primus, a, um	3 Πρῶτος, ου, m.
Two, 2, II.	Duo, æ, o	Δύο, δυοῖν
The Second	Secundus, a, um	3 Δεύτερος, έρου, m.
Three, 3, III.	Tres, tria, *com.*	Τρεῖς, τριῶν
The Third	Tertius, a, um	3 Τρίτος, ου, m.
Four, 4, IV.	Quatuor, *indec.*	Τέσσαρες, ρα, ων
The Fourth	Quartus, a, um	3 Τέταρτος, ου, m.
Five, 5, V.	Quinque, *indec.*	Πέντε
The Fifth	Quintus, a, um	3 Πέμπτος, ου, m.
Six, 6, VI.	Sex, *indec.*	Ἕξ
The Sixth	Sextus, a, um	3 Ἕκτος, ου, m.
Seven, 7, VII.	Septem, *indec.*	Ἑπτά
The Seventh	Septimus, a, um	3 Ἕβδομος, ου, m.
Eight, 8, VIII.	Octo, *indec.*	Ὀκτώ
The Eighth	Octavus, a, um	3 Ὄγδοος, όου, m.
Nine, 9, IX.	Novem, *indec.*	Ἐννέα
The Nineth	Nonus, a, um	3 Ἔνατος, ου, m.
Ten, 10, X.	Decem, *indec.*	Δέκα
The Tenth	Decimus, a, um	3 Δέκατος, ου, m.
Eleven, 11, XI.	Undecim, *indec.*	Ἕνδεκα
The Eleventh	Undecimus, a, um	3 Ἑνδέκατος, ου, m.
Twelve, 12, XII.	Duodecim, *indec.*	Δώδεκα
The Twelfth	Duodecimus, a, um	3 Δωδέκατος, ου, m.
Thirteen, 13, XIII.	Tredecim, *indec.*	Τρισκαίδεκα

The Thirteenth Decimus tertius. So the rest of the Ordinals from thirteen to twenty are compounded of *Decimus* and *Quartus*, *Quintus*, *&c.*

Fourteen, 14, XIV.	Quatuordecim, *indec.*	Δεκατέσσαρες & τεσσαρεσκαίδεκα
Fifteen, 15, XV.	Quindecim, *indec.*	Δεκαπέντε & πεντεκαίδεκα
Sixteen, 16, XVI.	Sexdecim vel Sedecim	Δεκαέξ & ἑκκαίδεκα
Seventeen, 17, XVII.	Septendecim	Δεκαέπτα & ἑπτακαίδεκα
Eighteen, 18, XVIII.	Octodecim	Δεκαοκτώ & ὀκτωκαίδεκα
Nineteen, 19, XIX.	Undeviginti	Δεκαεννέα

The Ordinals of these in Greek are, the Thirteenth Τρισκαιδέκατος, the Fourteenth Τεσσαρεσκαιδέκατος, &c.

Twenty,

Twenty, 20, XX.	Viginti	Εἴκοσι
Thirty, 30, XXX.	Triginta	Τριάκοντα
Fourty, 40, XL.	Quadraginta	Τεσσαράκοντα
Fifty, 50, L.	Quinquaginta	Πεντήκοντα
Sixty, 60, LX.	Sexaginta	Ἑξήκοντα
Seventy, 70, LXX.	Septuaginta	Ἑβδομήκοντα
Eighty, 80, LXXX.	Octoginta	Ὀγδοήκοντα
Ninety, 90, XC.	Nonaginta	Ἐννενήκοντα
An Hundred, 100, C.	Centum	Ἑκατόν

The Ordinals of these are, the Twentieth, *Vigesimus* or *vicesimus*, Εἰκοστὸς; the Thirtieth, *Trigesimus* or *tricesimus*, Τριακοστὸς; the Fourtieth, *Quadragesimus*, Τεσσαρακοστὸς, *&c.*

Two hundred, 200, CC.	Ducenti, æ, a	Διακόσιοι, αι, α
Three hundred, 300, CCC	Trecenti, æ, a	Τριακόσιοι, αι, α
Four hundr. 400, CCCC.	Quadringenti, æ, a	Τεσσαρακόσιοι, αι, α
Five hundred, 500, D.	Quingenti, æ, a	Πεντακόσιοι, αι, α, &c. ad eandem formam.
Six hundred, 600, DC.	Sexcenti, æ, a	
Seven hundr. 700, DCC.	Septingenti, æ, a	
Eight hund. 800, DCCC.	Octingenti, æ, a	
Nine hundred, 900, CM.	Nongenti, æ, a	
A Thousand, 1000, M.	Mille	Χίλιοι, αι, α
Ten thousand, 10000.	Decies mille	5 Μυριὰς, άδος, f.

FINIS.

ERRATA.

P*Aginâ* 35. *lineâ* 7. *pro* 5 'Αδὴν, ένος, m. *lege* 3 Χόνδρος, ε, m. *deinde supple* **A Glandule or Kernel**, Glandula, æ, *f.* 5 'Αδὴν, ένος, m. *Pag.* 40. *l.* 23. *pro* Rhoncus *lege* Rhonchus.

General Subject Index